AF577324

STATE OF NEW YORK
LICENSED PROFESSIONAL ENGINEER
EXCELSIOR

Energy Conservation in Wastewater Treatment Facilities

WEF Manual of Practice No. FD-2

Prepared by **Energy Conservation Task Force of the Water Environment Federation**
Thomas J. Kennedy, *Chair*

Dale J. Cagwin
Dals H. Chaudhary
James H. Clifton
Fred L. Hinker
David W. James
Franklin D. Munsey
Robert J. Ooten
Thomas J. O'Rourke
William F. Owen
Milton J. Pacheco
Lawrence J. Pakenas
Laurie Park
Frank H. Pearson
Beth Petrillo
David J. Reardon
Daniel F. Seidel
James L. Taylor
Rohit D. Trivedi
Thomas M. Walski
Katherine Van Zant
Dirk F. Young

Under the Direction of the
Municipal Subcommittee of the Technical Practice Committee

1997

Water Environment Federation
601 Wythe Street
Alexandria, VA 22314-1994 USA

IMPORTANT NOTICE

The material presented in this publication has been prepared in accordance with generally recognized engineering principles and practices and is for general information only. This information should not be used without first securing competent advice with respect to its suitability for any general or specific application.

The contents of this publication are not intended to be a standard of WEF and are not intended for use as a reference in purchase specifications, contracts, regulations, statutes, or any other legal document.

No reference made in this publication to any specific method, product, process, or service constitutes or implies an endorsement, recommendation, or warranty thereof by WEF.

WEF makes no representation or warranty of any kind, whether expressed or implied, concerning the accuracy, product, or process discussed in this publication and assumes no liability.

Anyone using this information assumes all liability arising from such use, including but not limited to infringement of any patent or patents.

Library of Congress Cataloging-in-Publication Data

Water Environment Federation. Energy Conservation Task Force.
Energy conservation in wastewater treatment facilities / prepared by Energy Conservation Task Force of the Water Environment Federation; under the direction of the Municipal Subcommittee of the Technical Practice Committee.

p. cm.—(Manual of practice. FD-2)

Includes bibliographical references (p.) and index.

ISBN 1-57278-034-7

1. Waterworks—Energy conservation. 2. Sewage disposal plants—Energy conservation. I. Water Environment Federation. Municipal Subcommittee. II. Title. III. Series.

TJ163.5.W36W38 1997

628.1—dc20 94-41952

CIP

Library of Congress Catalog Card No. 96-41952
ISBN 1-57278-034-7
Printed in the USA **1997**

Water Environment Federation

The Water Environment Federation is a not-for-profit technical educational organization that was founded in 1928. Its mission is to preserve and enhance the global water environment. Federation members are more than 42,000 water quality specialists from around the world, including environmental, civil, and chemical engineers, biologists, chemists, government officials, treatment plant managers and operators, laboratory technicians, college professors, researchers, students, and equipment manufacturers and distributors.

For information on membership, publications, and conferences contact

Water Environment Federation
601 Wythe Street
Alexandria, VA 22314-1994 USA
(703) 684-2400

Manuals of Practice

(As developed by the Water Environment Federation)

The Water Environment Federation (WEF) Technical Practice Committee (formerly the Committee on Sewage and Industrial Wastes Practice of the Federation of Sewage and Industrial Wastes Associations) was created by the Federation Board of Directors on October 11, 1941. The primary function of the committee is to originate and produce, through appropriate subcommittees, special publications dealing with technical aspects of the broad interests of the Federation. These manuals are intended to provide background information through a review of technical practices and detailed procedures that research and experience have shown to be functional and practical.

Water Environment Federation Technical Practice Committee Control Group

L.J. Glueckstein, *Chair*
T. Popowchak, *Vice-Chair*

G.T. Daigger
P.T. Karney
T.L. Krause
J. Semon
R. Zimmer

Authorized for Publication by the Board of Control
Water Environment Federation

Quincalee Brown, *Executive Director*

Acknowledgments

Principal Authors of the manual are

Dals H. Chaudhary
James H. Clifton
Fred L. Hinker
David W. James
Thomas J. Kennedy
Robert J. Ooten
Thomas J. O'Rourke
Frank H. Pearson
James L. Taylor
Rohit D. Trivedi
Thomas M. Walski
Dirk F. Young

In addition to the Task Force and Technical Practice Committee Control Group members, reviewers and contributors include

Franklin L. Burton
Robert V. Frey
Laurie Park

Authors' and reviewers' efforts were supported by the following organizations:

Andover Controls Corporation, Tampa, Florida
Azure Engineering, Inc., Keystone Heights, Florida
City of San Jose, California
County Sanitation Districts of Orange County, California, Fountain Valley
Department of Environmental Protection, New York, New York
Finkbeiner Pettis & Strout, Toledo, Ohio
Hazen and Sawyer, New York, New York
HDR Engineering, Inc., Folsom, California
Milwaukee Metropolitan Sewerage District, Milwaukee, Wisconsin
Montgomery Watson, Inc., Santa Rosa, New Mexico
New York State Energy Research and Development Authority, Albany
O'Rourke & Company, San Francisco, California
Owen Engineering & Management Consultants, Inc., Denver, Colorado
Texas Natural Resource Conservation Commission
University of California, Richmond
University of Florida, Gainesville
Westborough Wastewater Treatment Plant, Massachusetts
Wheelabrator EOS Inc., Hampton, New Hampshire
Wilkes University, Wilkes-Barre, Pennsylvania
Wyoming Valley Sanitation Authority, Wilkes-Barre, Pennsylvania

Federation technical staff project management was provided by J. Robert Schweinfurth.

Contents

List of Tables

List of Figures

Chapter 1
Energy-Efficient Wastewater Treatment

Introduction

This manual addresses energy and how it is used in the operation of wastewater treatment plants (WWTPs). Originally envisioned as a pocket book on energy conservation measures (ECMs) for wastewater treatment operators, the manual has been expanded to include basic concepts of energy as well as describing the technical basis for the energy required for various unit operations and unit processes commonly found in WWTPs. It is, therefore, an energy primer specific to wastewater treatment including helpful suggestions for controlling energy and energy costs. It is intended for WWTP managers and senior operators to provide a basis for greater understanding of the concepts of energy and efficient energy use to better manage WWTP energy consumption. It will also be useful to the design engineer in selecting energy-efficient equipment and processes.

Separate chapters discuss the major electricity-consuming equipment and systems found in WWTPs: motors, pumps, and aeration systems. The background information provided will aid in the selection of equipment in the design or the replacement process. Examples are shown to demonstrate how to compare differences in initial capital cost with operating cost changes because of differences in equipment efficiency. Another chapter is devoted entirely to solids-handling processes and discussions on energy recovery from anaerobic digestion systems and controlling energy losses from incinerators and sludge dryers.

The manual begins with a discussion of the basic technical concepts of energy describing the forms of energy, sources of energy, and conversion of energy between forms. It describes how utility companies charge for their services and details the various components of the utility bill. The manual concludes with ways to control electrical demand and ways to use alternate energy sources for greater overall economy.

Each of these chapters provides a theoretical discussion of the concepts of energy requirements and sources of inefficiency. Following many of the theoretical discussions are recommended ECMs for specific equipment or processes.

The manual also focuses on the overall costs of energy and how costs can be reduced through a better understanding of energy consumption, utility rate structures, and effective energy management. Utility rate structures are explained in understandable terms. Knowledgeable operators can often reduce their utility bills by taking advantage of time-shifting energy use and by controlling demand spikes.

Energy consumption is an important wastewater treatment system operation cost. Wastewater treatment plants are generally among the community's largest energy consumers (California Energy Commission, 1990), accounting for 1 to 2% (U.S. EPA, 1973) of a community's energy usage and 0.1 to 0.3% of the nation's total energy usage. Energy costs consume 15 to 30% of large WWTPs' and 30 to 40% of small WWTPs' operation and maintenance budgets. The energy cost of operating a WWTP continues to steadily rise because of the cost of fuels, inflation, and increasing wastewater discharge requirements. Advancing technology continually offers new opportunities for improving energy efficiency and the application of ECMs.

ROLE OF MANAGEMENT

Wastewater treatment plant management is responsible for establishing long-term objectives; developing supporting plans to accomplish the objectives; providing work direction, staff selection, and staff development; and ensuring overall facility performance. Management is responsible for operating the facilities in a cost-effective and safe manner while remaining in regulatory compliance.

Management sets the tone for energy conservation. Energy conservation programs should be ongoing. Committed management provides staff with leadership, encouragement, and resources to reduce energy consumption. The commitment to conserve energy must be supported and encouraged by utility management.

Operation and maintenance manuals should serve as information resources, and they should document any energy conservation features designed into a WWTP. For example, if variable-frequency drives (VFDs) have been installed on a pump as an ECM, the manual should provide information the operator or supervisor needs on the particulars of how to monitor the system energy use and how to knowledgeably operate the VFDs to maximize energy consumption. Data on pump system efficiency at various operating conditions should be presented on curves and charts in the manuals that tie to information on the field gauges, monitors, and computers. The manual should explain the design and provide basic design criteria.

It is important to keep operations and maintenance manuals up-to-date including any operating improvements developed by staff or consultants. Operations and maintenance manual updating has been made easier with the advent of computerization.

Management must have records of energy consumption of electricity, fuel, gas, and chemicals to review. The staff needs equipment that continuously monitors or records energy use and energy use changes.

The staff should be motivated to conserve "energy" and given work incentives such as payment for ideas that improve operating energy conservation and economy of operation. Pride programs that pay for implementable cost-saving ideas are used at many WWTPs.

In summary, management leads the way by defining energy conservation as a goal, motivates staff or hires consultants to develop workable plans to conserve, directs the plan's implementation, and provides the short- or long-term resources to implement and monitor progress.

IDENTIFYING ENERGY CONSERVATION MEASURES

In this manual, an ECM is defined as an operation or maintenance practice that leads to efficient use of energy. An ECM should either reduce the amount of energy consumed or reduce the price paid for energy. The procedure involved for identifying ECMs can be simple and does not necessarily have to involve energy management specialists, process engineers, or other experts. One procedure is as follows:

1. Review historical energy consumption (create 36-month spreadsheet trends) by the WWTP as a whole as well as any components that are individually metered,

2. Develop a list of the major energy-consuming equipment and conduct performance tests to determine the power draw and efficiency,
3. Define current process control procedures,
4. Analyze the data collected to identify ECMs,
5. Understand and become familiar with the procedures used by the power company to bill for energy used, and
6. Analyze and understand the utility bills and request time-of-day charts for kW and kWh used at the WWTP.

The WWTP's overall energy usage can be compared to published typical uses of electricity, fuel, or chemicals at other WWTPs. A simple but effective gauge is to determine the unit energy consumption in joules per cubic metre (J/m^3) or kilowatt-hours per million gallons (kWh/mil. gal) treated. Figure 1.1 summarizes typical energy consumption at a WWTP in the western U.S. As indicated, lagoon, trickling filter, and rotating biological contactor WWTPs are the most efficient, with the activated-sludge process and oxidation ditches being large energy consumers. Figure 1.1 assumes the WWTPs have influent pumping, primary clarification, anaerobic sludge digestion (except oxidation ditches), no effluent pumping, and no cogeneration. For example, a normal activated-sludge WWTP that consumes 2.5×10^6 J/m^3 (2 600 kWh/mil. gal) has potential energy savings that could amount to 40% or more when compared to other operating activated-sludge plants. However, if the activated-sludge WWTP operates at 1.1×10^6 J/m^3 (1 200 kWh/mil. gal), then the WWTP is efficient, and significant savings may be difficult to find.

General estimates of electricity used in WWTP unit processes for trickling filters, activated sludge, and advanced WWTPs without and with nitrification are shown in Tables 1.1, 1.2, 1.3, and 1.4, respectively. The actual energy used will vary at each WWTP, but the energy usages in these tables serve as guidelines.

The California legislature recognized, in 1988, that energy conservation opportunities exist at almost every WWTP. The California Energy Commission (1990) reported that WWTPs are often the single largest electricity users in a local government. As such, the commission targeted WWTPs for energy assistance programs. The programs provide grants to cities to identify opportunities for improvements in wastewater equipment that would reduce energy use or costs. The program funds WWTP audits and operator training. A $15 800 audit of the Eureka, California, WWTP and water system identified opportunities to improve pumping efficiencies. For capital costs of $56 800, the annual energy cost savings is $91 900 per year. The savings reduced the city's electric bill by 34%.

The commission found that operators were eager to implement hardware changes and reluctant to implement process recommendations. The commission is developing energy conservation schooling for operators to overcome this reluctance. The Orange County Sanitation District is another example of a

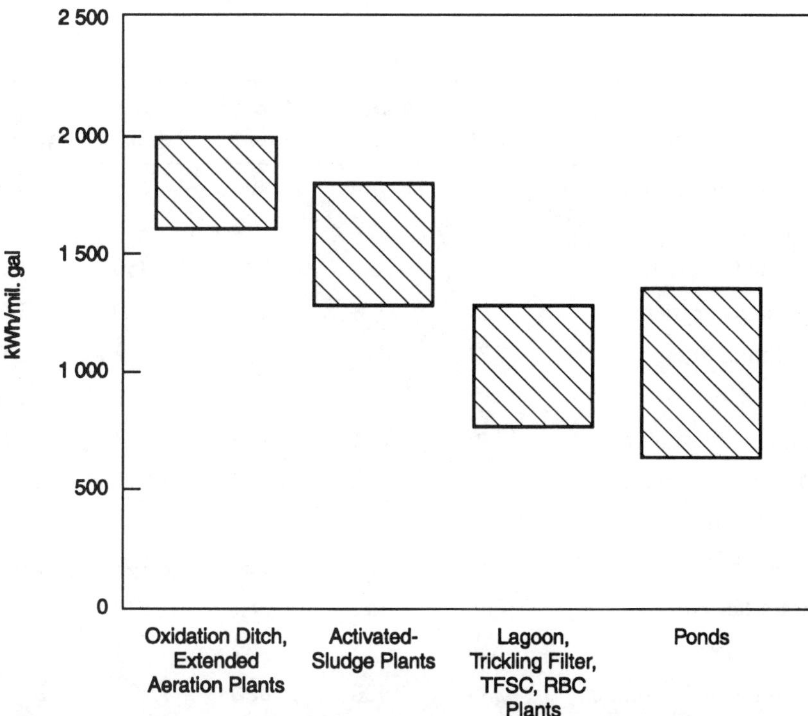

Figure 1.1 **Typical unit energy consumption of various types of wastewater treatment plants (TFSC = trickling filter solids contact and RBC = rotating biological contactor) (kWh/mil. gal × 951.1 = J/m^3)**

WWTP that recently implemented ECMs. The district has a history of energy conservation. In 1989, staff identified and put into action process changes and conservation techniques that resulted in savings of $650 000. The district has an energy conservation committee made up of staff and consultants that review energy conservation opportunities.

WHERE TO LOOK FOR ENERGY CONSERVATION MEASURES

It is useful to focus the initial search for ECMs in those areas where the greatest amounts of energy are consumed and, therefore, the greatest potential for savings exists. Most often, aeration and pumping systems will be the largest energy users, but at some WWTPs, other systems can use a large percentage of the electricity or energy.

Table 1.1 Electricity requirements for trickling filter wastewater treatment plants

Item	Electricity used, kWh/d[a] (except where noted)					
	1-mgd[b] plant	5-mgd plant	10-mgd plant	20-mgd plant	50-mgd plant	100-mgd plant
Wastewater pumping	171	716	1 402	2 559	6 030	11 818
Screens	2	2	2	3	6	11
Aerated grit removal	49	87	134	250	600	1 200
Primary clarifiers	15	78	155	310	776	1 551
Trickling filters[c]	352	1 319	2 528	4 686	11 551	22 826
Secondary clarifiers	15	78	155	310	776	1 551
Gravity thickening	6	15	25	37	75	138
Dissolved air flotation	na[d]	na	1 805	2 918	6 257	11 819
Aerobic digestion	1 000	2 000	na	na	na	na
Anaerobic digestion	na	na	1 100	2 100	5 000	11 000
Belt filter press	na	192	384	579	1 164	2 139
Chlorination	1	5	27	53	133	266
Lighting and buildings	200	400	800	1 200	2 000	3 000
Totals	1 811	4 892	8 517	15 005	34 368	67 319
Average flow rate, mgd	1	5	10	20	50	100
Unit electricity use, kWh/mil. gal[e]	1 811	978	852	750	687	673
Energy recovery (from biogas combustion)	na	na	2 800	5 600	14 000	28 000
Net consumption[f]	1 811	4 892	5 717	9 405	20 368	39 319
Unit net electricity use, kWh/mil. gal	1 811	978	572	470	407	393

[a] kWh/d × 41.67 = W.
[b] mgd × (4.383×10^{-2}) = m^3/s.
[c] With recirculation pumping.
[d] Not applicable.
[e] kWh/mil. gal × 951.1 = J/m^3.
[f] Total less electricity from energy recovery.

Table 1.2 **Electricity requirements for activated-sludge wastewater treatment plants**

Item	Electricity used, kWh/d[a] (except where noted)					
	1-mgd[b] plant	5-mgd plant	10-mgd plant	20-mgd plant	50-mgd plant	100-mgd plant
Wastewater pumping	171	716	1 402	2 559	6 030	11 818
Screens	2	2	2	3	6	11
Aerated grit removal	49	87	134	250	600	1 200
Primary clarifiers	15	78	155	310	776	1 551
Aeration (diffused air)	532	2 660	5 320	10 640	26 600	53 200
Return sludge pumping	45	213	423	724	1 627	3 131
Secondary clarifiers	15	78	155	310	776	1 551
Gravity thickening	6	15	25	37	75	138
Dissolved air flotation	na[c]	na	1 805	2 918	6 257	11 819
Aerobic digestion	1 200	2 400	na	na	na	na
Anaerobic digestion	na	na	1 400	2 700	6 500	13 000
Belt filter press	na	192	384	579	1 164	2 139
Chlorination	1	5	27	53	133	266
Lighting and buildings	200	400	800	1 200	2 000	3 000
Totals	2 236	6 846	12 032	22 283	52 544	102 824
Average flow rate, mgd	1	5	10	20	50	100
Unit electricity use, kWh/mil. gal[d]	2 236	1 369	1 203	1 114	1 051	1 028
Energy recovery (from biogas combustion)	na	na	3 500	7 000	17 500	35 000
Net consumption[e]	2 236	6 779	8 532	15 283	35 044	67 824
Unit net electricity use, kWh/mil. gal	2 236	1 356	853	764	701	678

[a] kWh/d × 41.67 = W.
[b] mgd × (4.383 × 10^{-2}) = m^3/s.
[c] Not applicable.
[d] kWh/mil. gal × 951.1 = J/m^3.
[e] Total less electricity from energy recovery.

Table 1.3 Electricity requirements for advanced wastewater treatment plants without nitrification

Item	Electricity use, kWh/d[a] (except where noted)					
	1-mgd[b] plant	5-mgd plant	10-mgd plant	20-mgd plant	50-mgd plant	100-mgd plant
Wastewater pumping	171	716	1 402	2 559	6 030	11 818
Screens	2	2	2	3	6	11
Aerated grit removal	49	87	134	250	600	1 200
Primary clarifiers	15	78	155	310	776	1 551
Aeration (diffused air)	532	2 660	5 320	10 640	26 600	53 200
Return sludge pumping	45	213	423	724	627	3 131
Secondary clarifiers	15	78	155	310	776	1 551
Chemical addition	80	290	552	954	2 187	4 159
Filter feed pumping	143	445	822	1 645	3 440	6 712
Filtration	137	247	385	709	1 679	3 295
Gravity thickening	6	15	25	37	75	138
Dissolved air flotation	na[c]	na	2 022	3 268	7 008	13 237
Aerobic digestion	1 200	2 400	na	na	na	na
Anaerobic digestion	na	na	1 400	2 700	6 500	13 000
Belt filter press	na	228	457	689	1 385	2 545
Chlorination	1	5	27	53	133	266
Lighting and buildings	200	400	800	1 200	2 000	3 000
Totals	2 596	7 864	14 081	26 051	60 822	118 814
Average flow rate, mgd	1	5	10	20	50	100
Unit electricity use, kWh/mil. gal[d]	2 596	1 573	1 408	1 303	1 216	1 188
Energy recovery (from biogas combustion)	na	na	3 500	7 000	17 500	35 000
Net consumption[e]	2 596	7 964	10 581	19 051	43 322	83 814
Unit net electricity use, kWh/mil. gal	2 596	1 573	1 058	953	866	838

[a] kWh/d × 41.67 = W.
[b] mgd × (4.383 × 10^{-2}) = m^3/s.
[c] Not applicable.
[d] kWh/mil. gal × 951.1 = J/m^3.
[e] Total less electricity from energy recovery.

Table 1.4 Electricity requirements for advanced wastewater treatment plants with nitrification

Item	Electricity use, kWh/d[a] (except where noted)					
	1-mgd[b] plant	5-mgd plant	10-mgd plant	20-mgd plant	50-mgd plant	100-mgd plant
Wastewater pumping	171	716	1 402	2 559	6 030	11 818
Screens	2	2	2	3	6	11
Aerated grit removal	49	87	134	250	600	1 200
Primary clarifiers	15	78	155	310	776	1 551
Aeration (diffused air)	532	2 660	5 320	10 640	26 600	53 200
Biological nitrification	346	1 724	3 446	6 818	16 936	33 800
Return sludge pumping	54	256	508	869	1 952	3 757
Secondary clarifiers	15	78	155	310	776	1 551
Chemical addition	80	290	552	954	2 187	4 159
Filter feed pumping	143	445	822	1 645	3 440	6 712
Filtration	137	247	385	709	1 679	3 295
Gravity thickening	6	15	25	37	75	138
Dissolved air flotation	na[c]	na	2 022	3 268	7 008	13 237
Aerobic digestion	1 200	2 400	na	na	na	na
Anaerobic digestion	na	na	1 700	3 200	7 800	15 600
Belt filter press	na	228	457	689	1 385	2 545
Chlorination	1	5	27	53	133	266
Lighting and buildings	200	400	800	1 200	2 000	3 000
Totals	2 951	9 631	17 912	33 514	79 383	155 540
Average flow rate, mgd	1	5	10	20	50	100
Unit electricity use, kWh/mil. gal[d]	2 951	1 926	1 791	1 676	1 588	1 558
Energy recovery (from biogas combustion)	na	na	3 500	7 000	17 500	35 000
Net consumption[e]	2 951	9 631	14 412	26 514	61 883	120 840
Unit net electricity use, kWh/mil. gal	2 951	1 926	1 441	1 326	1 238	1 208

[a] kWh/d × 41.67 = W.
[b] mgd × (4.383×10^{-2}) = m^3/s.
[c] Not applicable.
[d] kWh/mil. gal × 951.1 = J/m^3.
[e] Total less electricity from energy recovery.

Advanced treatment processes require relatively large amounts of energy, much of which is embodied in chemicals added for purposes such as phosphorus precipitation, refractory organic removal, and desalinization.

RANKING AND IMPLEMENTATION OF ENERGY CONSERVATION MEASURES

Proper ranking of ECMs is crucial to the success of the energy conservation program. Because the goal of the program is to reduce costs, it is logical that economic benefit should be the most important criterion in assigning priorities. Other factors that should be considered include cost of implementation, time required to recover this cost, and probability that a particular ECM will achieve its projected savings.

Energy conservation measures may be divided into three categories: low cost, moderate cost, and high cost. The actual dollar amounts assigned to these categories will vary with WWTP size and financial resources.

The first group includes actions requiring little change in operational routine. Generally, the funds needed for implementation, if any, can come out of the WWTP expense budget. As a rule, these ECMs should be implemented at the earliest possible date.

The moderate-cost group may require staff evaluations, minor drawings or specifications, outside contractor assistance, and budget item procurement of labor and materials. Energy conservation measures in this group can be readily compared and prioritized on the basis of simple payback periods. The simple payback period is calculated by dividing the cost of implementing the ECM by the projected monthly or yearly savings it will provide. Energy conservation measures with the shortest payback periods should be given the highest priority.

Energy conservation measures in the last group will probably require evaluation by the city engineering staff or outside engineering consultant. Evaluation and implementation would probably be separate budget line items. Major structural rehabilitation, automation systems, and major equipment change-outs are examples of actions in the group. Economic analyses that take into account the time value of money are required to compare alternative high-cost ECMs and justify their implementation.

REFERENCES

California Energy Commission (1990) *The Second Report to the Legislature on Programs Funded Through Senate Bill 880*. PL 400-89-006, Sacramento.

U.S. Environmental Protection Agency (1973) *Electrical Power Consumption for Municipal Wastewater Treatment*. EPA-R2/73-281, Washington, D.C.

SUGGESTED READING

U.S. Environmental Protection Agency (1978) *Energy Conservation in Municipal Wastewater Treatment*. EPA-430/94-77-011, Washington, D.C.

Chapter 2
Energy Basics

Energy is one of those terms that is familiar to everyone but seems to have many different interpretations regarding its meaning. Aside from personal or physical energy, the energy that is the topic of this manual is energy as defined by the science of physics. The technical definition of energy most commonly used is *the ability or capacity to do work.* Part of the problem with defining and understanding energy is that there are many types of energy. These types can be divided into two broad classifications: potential energy and kinetic energy. Most people think of energy in the kinetic sense. Kinetic energy is energy that is moving or is in actual production or use. Potential energy, conversely, is stored energy. It is the energy contained within fuel oil, for example, or in an elevated water tank or reservoir that can produce kinetic energy (that is, has the ability to produce work) when released. Energy is, among other things, the fuel in the ground as well as the electricity entering homes.

The most interesting aspect of energy is that it is represented in many different forms and can be changed from one form to another. Common forms of energy are chemical, electrical, mechanical, thermal, radiant, and nuclear. Most people are familiar with the conversion of fossil fuel through burning (chemical energy) to form steam (thermal energy), which drives a turbine (mechanical energy) to rotate a generator and form electricity (electrical energy). While most forms of energy are interchangeable, a certain amount of efficiency is lost in the conversion from one form to another. For example, the efficiency of conversion in the last example from fossil fuel to electricity is less than 40% in high-temperature units and less than 30% in boilers operating at conventional temperature. Most of the efficiency loss is owing to the loss of energy in the form of unusable heat. Cogeneration is an attempt to make use of some of the heat that is otherwise wasted in electricity production. *Cogeneration* is a term used when some of the heat from energy production is beneficially used to make steam or provide space heating. Heat from internal-combustion engines is especially usable because of the relatively high rejection temperatures of the air and cooling water. Overall energy use in cogeneration systems can exceed 80%. Losses are inherent in any energy conversion system. When converting a liquid fuel, for example, to heat a building by converting #2 fuel oil to heat, some energy exits through the chimney as hot exhaust gas. Only a certain percentage of the heat of the flame and air in the firebox can be captured by a heat exchanger and delivered throughout the building. Typically, this amount ranges from 82 to 85%. From an energy perspective, then, it would make much more sense to use fossil fuels for heating purposes than to use electrical space heaters or heat pumps because most furnaces are 80% efficient or better. Even a 100% efficient electric heater can only claim the 25 to 40% efficiency conversion from the fossil fuel.

When a gas such as air is moved through a blower or liquid is moved through a centrifugal pump, some of the energy input is converted to heat because of slippage of the fluid between the stationary and rotating parts of the device. A certain amount of slippage is designed into the equipment because that loss is necessary for the rotating element to spin freely. Excess slippage caused by equipment wear is an unnecessary loss of efficiency and can be limited, just as loss of furnace efficiency can be limited by ensuring that the proper ratio of fuel to air is maintained and heat exchange surfaces are kept clean.

BERNOULLI EQUATION

One of the basic laws of physics is the law of conservation of energy, which states that energy can be neither created nor destroyed. Consider, for example, water flowing in a single pipeline with only one intake and one discharge and without energy input or output. The law of conservation of energy states that the energy at any point in the pipeline is the same as at any other point in the pipeline, adjusted for any energy losses between them. Bernoulli developed an equation to account for all energy terms in a closed conduit. He expressed all energy in the flowing conduit in terms of *head,* which is typically expressed as energy per unit weight (for example, newton metres per newton equals metres, or foot-pounds per pound equals feet, generally expressed in the metric system as metres of water, or simply metres, and in the English system as feet). The equation states that the sum of the velocity head, pressure head, and elevation head at any point in the conduit is the same as the sum of the velocity head, pressure head, and elevation head at any other point, adjusted for head loss between the two points.

$$\frac{V_1^2}{2g} + \frac{p_1}{\rho} + Z_1 = \frac{V_2^2}{2g} + \frac{p_2}{\rho} + Z_2 + H_L \tag{2.1}$$

Where

$\frac{V^2}{2g}$ = velocity head, m (V = m/s and g = m/s^2)—the kinetic energy of the flowing liquid (from physics, kinetic energy [ke] = ½ mv^2, weight [w] = $m \times g$, and kinetic energy per unit weight [ke/w] = $mv^2/2mg = v^2/2g$, where w = weight of fluid, m = mass of fluid, v = velocity of fluid, and g = acceleration caused by gravity).

$\frac{p}{\rho}$ = pressure head, m (p = kg/m^2 and ρ = kg/mL)—the gauge pressure at the point of measurement. It accounts for static head above or suction head below point of measurement plus induced pressure from valve closure or compressible gas cushion in the system. Because head is measured in metres or feet of the fluid, one must account for the density (ρ) of the fluid in converting from units of pressure to head to arrive at the proper pressure head.

Z = elevation head, m—the amount of energy that the fluid possesses at the point of measurement because of its elevation measured from a baseline. For a point in a system with a pressure gauge, the elevation head is the potential energy as measured by the distance from the baseline (usually below) to the pressure gauge.

Pressure head and elevation head are often confused. Pressure head can be measured by a gauge and includes both pressure induced in the system, if any, and elevation of the fluid above the point of measurement. The latter is referred to as *static head.* Elevation head measures the distance from an arbitrary baseline (usually below the measurement point) to the point of measurement.

Head loss refers to frictional losses. Frictional losses are a function of the length of pipe (plus equivalent length of fittings), diameter of pipe, pipe roughness, fluid viscosity, and fluid velocity. Head loss is approximately related to the direct square of the velocity and linearly related to other factors, so it increases significantly as flow increases. Pipes are typically designed to have velocities between 1.2 and 3 m/s (4 and 10 ft/sec) at normal flow. At greater than 3 m/s (10 ft/sec), head losses become excessive.

ENERGY USE IN WASTEWATER TREATMENT PLANTS

The major consumers of energy in a wastewater treatment plant (WWTP) are pumps, blowers, mechanical aerators, and solids-handling systems.

PUMPING. To determine the power requirements for a pump, it is necessary to know the amount of liquid to be pumped and the head that it must overcome. The pump must elevate the water from the lower reservoir to the higher reservoir, must provide pressure if necessary in the system, and must overcome all friction losses through the line as well as through the pump, valves, and fittings. The sum of these heads is used for pump design and is called the *total dynamic head* (TDH). Recall that head was defined as "energy per unit weight" for the units of metres or feet.

The TDH is derived from the Bernoulli equation using the difference between the point of maximum energy that the pump must overcome and the point of minimum energy from which the pump draws. Generally, these two points are the surfaces of the fluid on the discharge side and intake side of the pump. The difference in energy calculated by this form of the Bernoulli equation is equivalent to the energy that must be supplied by the pump. Total dynamic head represents the effective distance that the weight (force) of the fluid must move. The product of flow and TDH is directly related to the horsepower required at the pump output.

Because most pumping applications involve pumping from one open reservoir to another, the Bernoulli equation can be simplified to the difference in elevations between the water surfaces of the two reservoirs plus the friction losses between them. Open reservoirs would have no differential pressure at the water surfaces and almost no velocity head in most cases. The Bernoulli equation then becomes

$$\text{TDH} = \left(\frac{V_d^2}{2g} - \frac{V_i^2}{2g}\right) + \left(\frac{p_d}{\rho} - \frac{p_i}{\rho}\right) + (Z_d - Z_i) + H_L \tag{2.2}$$

Which simplifies for open reservoirs to

$$\text{TDH} = (Z_d - Z_i) + H_L \tag{2.3}$$

Where

Z_d	=	discharge surface elevation, m; and
Z_i	=	intake elevation of the fluid, m.

With a reference elevation of the centerline of the pump, TDH is often expressed as

$$\text{TDH} = (Hd - Hs) + Hf \tag{2.4}$$

Where

Hd	=	discharge head, m;
Hs	=	suction head, m; and
Hf	=	friction head losses.

The static head (difference in elevation heads) represents the vertical distance between the liquid level from which the pump will *suck* and the highest liquid level to which the pump will discharge. It is normally the largest component of TDH. However, in systems with either small-diameter piping or long pipelines, friction losses can become much greater than static head.

Friction head is the result of friction losses because of pipe length, pipe diameter, pipe roughness, and all fittings and restrictions in the pipeline. Friction head increases exponentially with increasing velocity in pipes and fittings. Friction head for a 25-mm (1-in.) diameter, smooth-interior pipe at a water velocity of 3 m/s (10 ft/sec) is 12 m per 30 m (40 ft per 100 ft) of (equivalent length) straight pipe.

In sizing a pump, a design engineer starts with the required flow and then calculates the TDH for the maximum pump flow. Maximum *system* condition is also calculated if more than one pump is used on the same pipe system. Knowing the desired pump capacity and maximum (or design) TDH, the minimum or *theoretical* motor horsepower (horsepower × 745.7 = watts) for the pump can be calculated easily.

To elevate the liquid and overcome piping resistance, the pump system must perform *work.* Work, a form of energy commonly measured in newton metres (n·m) or foot-pounds (ft-lb), is defined as moving a force through a distance. In pumping, the weight of water is the force and the TDH is the distance. Power, which is often measured in watts (W) or horsepower (hp), is defined as work per unit time: 1 hp (745 W) is equivalent to 33 000 ft-lb/min. Volume or flow can easily be converted to weight terms.

For example, to pump 1 mgd, or 694 gpm, of water at a conventional TDH of 33 ft, the following equation would apply (see Appendix for conversions from English to metric units):

$$694 \text{ gpm/mgd} \times 8.34 \text{ lb/gal} \times 33 \text{ ft} = 191\,000 \text{ ft-lb/min/mgd} \tag{2.5}$$

Converting to horsepower using the above conversion factor, the pump output power, also known as *water* horsepower, can be calculated as follows:

$$\frac{191\ 000 \text{ ft-lb/min/mgd}}{33\ 000 \text{ ft-lb/min/hp}} = 5.79 \text{ hp/mgd} \tag{2.6}$$

Centrifugal pumps have efficiencies of approximately 70 to 75%. Using a pump efficiency of 74% for a 1-mgd pump, the actual power required for pump input, also known as *brake* horsepower, would be

$$\frac{5.79 \text{ hp}}{0.74 \text{ efficiency}} = 7.82 \text{ hp} \tag{2.7}$$

Because motors are only manufactured at nominal sizes, a 10-hp (7 500-W) motor would be chosen for this application. The next smaller size motor, 7.5 hp (5 500 W), is too small.

Motors are rated by their output (shaft-hp) rather than by their input, and they are approximately 75 to 95% efficient (see Chapter 4). A 10-hp (7 500-W), high-efficiency motor operating at 75% load would be approximately 88% efficient. Therefore, the power to the motor, also known as *wire* horsepower, would be

$$\frac{7.82 \text{ hp}}{0.88 \text{ efficiency}} = 8.89 \text{ hp} \tag{2.8}$$

Because there are 0.746 kW/hp, the kilowatt draw for an unthrottled, 1-mgd (3 800-m^3/d), fixed-speed pump operating at 33 ft (10 m) TDH and with the pump and motor efficiencies specified would be

$$8.89 \text{ hp/mgd} \times 0.746 \text{ kW/hp} = 6.63 \text{ kW/mgd} \tag{2.9}$$

It is important to note that the motor for the pump operating at the design conditions would be drawing only 8.89 hp (6 600 W) of line electricity and not the full 11.4-hp (8 500-W) input rating of the 10-hp (7 500-W) motor. Similarly, if the pump is throttled, it will draw less energy because it will do less work. The reduction in energy requirements from throttling are not entirely proportional to the flow reduction because a throttled pump operates less efficiently.

A variable-speed pump must include a correction for the transmission inefficiency. Variable-speed transmission devices vary in efficiency from 50 to 95%. Magnetic and slip clutch devices are the most inefficient at 50 to 60% efficiency; gear boxes and sheaves are 70 to 80% efficient; and electronic drives (such as variable-frequency drives) are 85 to 95% efficient. Even the most efficient variable-speed pump is not as efficient as a direct-drive pump.

The following are energy conservation measures (ECMs) to consider:

- Pumping efficiency is not easy to improve, so preventing original pump and motor efficiencies from deteriorating is critical.
- For greatest economy in a multiple-pump situation, use the most efficient pumps for normal flows and the least efficient pumps for peak flows.
- Use fixed-speed rather than variable-speed pumps when the downstream process can handle the abrupt flow changes. When a variable-speed system is required, select energy-efficient drives and pumps equipped with energy-efficient motors over those with conventional motors.
- Minimize the number of pumps operating at the same time, so that they do not compete with each other for energy.
- Keep pumps and motors properly maintained and operating close to their design efficiencies. Check for and replace worn wear rings, which can significantly reduce pump efficiency. Check impellers and rebuild as necessary.

POWER

Power is generally defined as the rate at which work is performed. It is the speed at which the work is done or the rate at which energy is expended. The size of a motor or engine is defined not by the total amount of work to be done, but by the rate at which it can do work. The basic unit of power in the metric system is joules per second (J/sec), more commonly known as watts. In the English system, some of the common units of measure are foot-pounds per second, horsepower (1 hp = 550 ft-lb/sec), and British thermal units per hour (Btu/hr). Electric motor power is commonly measured in horsepower or kilowatts (watts); boiler and air conditioner power are measured in Btu/hr (watts).

The power applied multiplied by the time it is applied represents total energy consumption.

$$\text{Energy (kWh)} = \text{power (kW)} \times \text{time (hr)} \tag{2.10}$$

$$\text{Energy (hp-hr)} = \text{power (hp)} \times \text{time (hr)} \tag{2.11}$$

$$\text{Energy (Btu)} = \text{power (Btu/hr)} \times \text{time (hr)} \tag{2.12}$$

$$\text{Energy (ft-lb)} = \text{power (ft-lb/sec)} \times \text{time (sec)} \tag{2.13}$$

Note that watt and horsepower do not appear to be rate units, but they are. All power units are measured as work per unit time.

Energy consumption is generally measured in British thermal units (Btu) (kilojoules [kJ]) if consumed as a fuel or kilowatt hours (kWh) if consumed as electricity. Natural gas contains approximately 1 000 Btu/cu ft (37 000 kJ/m^3), and billing units are typically 100 cu ft (2.8 m^3) of gas measured at standard temperature and pressure conditions, 100 000 Btu (otherwise known as *therms*) (105 000 kJ), or million Btu (MB). Fuel oil characteristically contains 140 000 Btu per gallon (3.9×10^7 kJ/m^3), and consumption is measured as gallons (cubic metres). Table 2.1 provides conversion factors so that energy consumption measured by one method may be changed to different units. In addition, the Appendix provides metric conversions for English units used in this manual.

Table 2.1 Conversion chart for various forms and units of energy

	Work		Heat		Electric	
Unit	**ft-lb**	**kg·m**	**Btu**	**kcal**	**hp-hr**	**kWh**
ft-lb	1	0.138 3	1.286×10^{-3}	3.241×10^{-4}	5.050×10^{-7}	3.766×10^{-7}
kg-m	7.231	1	9.302×10^{-3}	2.343×10^{-3}	3.654×10^{-6}	2.724×10^{-6}
Btu	777.9	107.5	1	0.252 0	3.927×10^{-4}	2.928×10^{-4}
kcal	3 086	426.8	3.968	1	1.558×10^{-3}	1.162×10^{-3}
hp-hr	1.980×10^6	2.737×10^5	2 547	641.7	1	0.745 7
kWh	2.655×10^6	3.671×10^5	3 415	860.5	1.341	1

ELECTRICITY

Electricity is a condition in which electrons are displaced from atoms. Because it takes energy to displace the electrons, the displaced electrons represent energy that can do useful work. Therefore, electricity is a form of energy that results from an unbalanced atomic condition. If the electrons are merely displaced, a condition of static electricity exists. If the electrons flow back to the atoms, a condition of current electricity exists during the time of movement. Electricity can be generated in several ways, but some other form of energy is required, and the other energy form must be converted to electricity. The most commonly used methods to generate electricity are magnetic (for example, generators and alternators), chemical (for example, batteries), and photocells (for example, light sensors and solar panels), with magnetic generation being the major source of artificially produced electricity in the world. Fossil fuels, hydropower, geothermal energy, nuclear energy, wind, and tidal energy are used to develop the mechanical energy to magnetically produce electricity.

In any electrical circuit at least four factors are present: electrical pressure (voltage), electron flow or current (amperage), resistance, and power (wattage). Voltage, current, and resistance are related through Ohm's law in the following equation:

$$E = IR \tag{2.14}$$

Where

E	=	electrical pressure, volts (V);
I	=	electrical current, amperes (A); and
R	=	resistance, ohms.

Ohm's law does not apply to alternating current (ac) circuits, which contain coils or capacitors. Another relationship exists for simple circuit power P, in which

$$P = EI \tag{2.15}$$

Power for three-phase circuits is discussed in more detail along with power factor in Chapter 3.

ALTERNATING CURRENT VOLTAGE. Alternating current voltage is determined by its effectiveness or the power it produces compared to direct current (dc) voltage. Because the voltage of each alternating current starts at zero, increases to maximum, then decreases to zero, ac voltage is continually varying and reversing and is never at a constant value. Therefore, an effective value has to be determined. When the voltage of an alternation reaches peak, for example 90 deg, it is known as *peak* or *wave-crest* voltage. The effective voltage, or working voltage, of ac is 0.707 times the value of the peak voltage. This value was determined by comparing ac voltage with dc voltage in producing heat in a resistor. From a point of effectiveness, 1 ac-effective V (0.707 of the peak ac voltage) is equal to 1 dc V. Effective voltage is also known as *root-mean-square* (rms) voltage. All common electrical instruments have readouts that are in effective or rms voltage (for example, 120 V is the effective voltage and 170 V is the peak voltage of common ac current).

VOLTAGE DROP. Voltage is electrical pressure. In any series electrical circuit, the total volts applied to the circuit are entirely used. No pressure returns to the generator. All amperes that leave the generator return, but volts are completely spent in the circuit. If ammeter readings are taken around a circuit, it will be found that the ampere readings are all the same. The volts spent or dropped in a part of a circuit are in direct proportion to the resistance in that part of the circuit.

Copper is the most widely used electrical conductor because of its high electrical transfer efficiency and relatively low cost compared with precious metals, which have higher transfer efficiencies. A voltage drop is present in all circuits because of resistance in the conductor. Resistance increases directly with length of wire and inversely with the diameter of the wire.

Resistance caused by improper wire sizes, poor wire connections, or rough contact points in switches results in a voltage drop of the line voltage, and a buildup of heat and loss of power. For example, in a simple circuit with a device that requires 100 V and 5 A and has a 1-ohm resistance because of the wire, the line voltage needed will be 105 V because the 1-ohm resistance at 5 A will require 5 V of line pressure to overcome the resistance in the wire. Additionally, the power loss is represented by an equation generated by combining Equations 2.14 and 2.15 above, as follows:

$$P = I^2R \tag{2.16}$$

For this example, the power loss is 25 W ($5^2 \times 1$). The 25-W power loss is energy lost in the form of heat dissipated over the length of the wire.

Regardless of the form of the resistance, heat will be produced when electricity flows through it. This is true of connections that may loosen from the constant warming and cooling of the electric circuits, which causes expansion and contraction of circuit metals, or of contact points in switches, which wear from use. Loose connections, worn contact points, or other undesirable resistances can be found using infrared thermography or simple voltage measurements across the connection or switch. Infrared photographs or video cameras make heat visible, with greater temperatures producing brighter images. As previously discussed, circuit resistance results in a voltage drop. A simple voltmeter reading across a connector or switch from time to time will show whether resistance is increasing or at excessive levels.

The following ECMs are recommended:

- Connections and switches on all major WWTP power-driven equipment, buses, and transformers should be checked at least once per year so that simple corrective action can be taken before serious problems with equipment develop or significant power losses occur.
- Motors should be operated as close to nameplate voltage as practical. Motors are generally incapable of operating on voltages varying more than 10% from the nameplate. Any deviation from the nameplate rating affects the motor's efficiency. In general, it is recommended that the motor line drop never exceed 5% of the line voltage.

Chapter 3 Utility Billing Procedures and Incentives

INTRODUCTION

Utility rate structures have no direct effect on energy efficiency or energy use, although high charges may provide incentives to look for ways to reduce the wastewater treatment plant's (WWTP's) utility costs and to implement energy conservation measures. There are many opportunities to save money by understanding not only the current rate structure being applied to current billing but also knowing what to look for with respect to alternate billing structures that may be available. There may be a variety of electric rate schedules and special contracts available in addition to the standard industrial tariff, especially for large municipal WWTPs. Specific terms and conditions of such special rates and contracts will vary widely from one utility to another and between states. Such special rates and contracts are designed to attain the load distribution and other management objectives of the utility serving the area. Additionally, the utility may be under federal or state mandates to encourage users to practice energy conservation or load-shifting that may reduce rates or provide rebates.

Deregulation of utilities may offer additional opportunities for large consumers who can take advantage of less expensive nonlocal suppliers. Like the deregulation of the phone company, natural gas is already being provided by nonlocal suppliers through local supplier transmission mains. Similarly, the Energy Policy Act of 1992 has opened up the channels for supply of electricity by nonlocal providers at the wholesale level.

Implementation at the retail level, known as *retail wheeling,* is up to each state because regulations must be approved by state utility regulatory commissions. Serious issues, however, must be overcome before retail wheeling becomes widely available. Not only do rates vary widely from state to state, but utilities also have a significant number of power plants that are not efficient enough to compete in an open market, especially against newer plants that use efficient, combine-cycle gas turbines.

In this chapter, we will present a broad overview of the types of utility charges along with cost savings opportunities offered by local utilities and key factors that can influence selection of an appropriate rate or contract option. Electric utility companies are service providers, unlike fuel companies, which are product suppliers. Once a fuel is purchased by a plant, all of the factors that influence and affect its efficiency are controlled by the user. On the other hand, the way electricity is consumed does affect the utility, and its rates are structured to account for this. Any rate or rate component may be structured in a variety of ways.

Electric Rates

BILLING CHARGES. The billing for electric service can take many forms depending on the economic objectives and administrative concerns that are being addressed by the utility. Unlike an electric invoice for home use, which may only contain a customer charge, energy charge, and taxes, WWTPs typically are subject to complex rate structures classified under the heading *industrial use*. Municipalities will often have rate classifications separate from industry, but, nonetheless, they are similar to industrial rate classifications. In general, the monthly invoice for all industrial consumers will contain the following minimum charge categories: customer charge, energy charge, demand charge, power factor (PF) penalty, fuel cost adjustment, and taxes (note that most municipal WWTPs are exempt from state and local taxes). Additionally, there are many other charges and surcharges that may be added to the bill or credits given that may be deducted. Rates for energy and demand are normally specified in use brackets much like income taxes but with lower rates for each bracket of increased use. However, utilities are beginning to implement increased rates for increased use as an incentive to reduce consumption. Home bills, for example, may have a reduced rate for the first bracket of use, which is often referred to as a *lifeline* rate. Increased use above the quantities allowed by the lifeline amount typically leads to substantially higher rates.

Customer Charge. The customer charge is a fee structured to compensate the utility for administrative costs incurred in servicing the customer (for example, for reading meters, preparing and mailing bills, and collections). Customer charges for WWTPs are lower than energy and demand charges. They are often less than $100 per month but can be higher depending on the sophistication of the meter used.

Energy Charge. The energy charge is the charge for the actual energy used, measured in kilowatt-hours (kWh) (megajoules [MJ]), during the billing period. It represents the utility company's operating cost for generating and supplying the electricity, including its profits. Following the energy crises of the 1970s and the subsequent volatility in the prices of fuels, utilities elected to modify standard billing rate schedules through enactment of a monthly adjustment to compensate for the volatile price of fuels. This has allowed them to keep published energy charges fairly constant, while adjusting for their actual costs of fuel on a monthly basis through a fuel cost adjustment charge.

While the utility's cost for generation of electricity does not necessarily depend on the time of day, utilities have elected to assess premium charges during peak usage periods for the community to discourage energy use during these peak periods. Lower costs for nonpeak periods are also offered to encourage users to shift the time of their usage. This is a utility's effort to balance its production schedules and maximize use of existing facilities. Peak periods of usage in many parts of the U.S. occur during the hottest days of the summer, when air conditioning demand is greatest. Utilities must strain their resources to deliver the energy demanded by their customers during these brief peak demand periods, while having much of their expensive equipment idle most other times. Ironically, a utility's actual cost for generating electricity, because of greater efficiency and relatively flat labor cost, is lowest during peak periods, but it is charged at the highest rate.

Energy use is measured using a wattmeter, which consists of a voltmeter and ammeter making concurrent measurements. Older units are electromechanical, with the familiar hard-to-read clock-type dials, while modern units are electronic and have digital readouts.

Demand Charge. The demand charge represents the maximum power drawn during the billing period (usually averaged over a contiguous 15-minute or longer period). It is expressed in kilowatts as the average power demanded during that peak period. Power spikes from in-rush current when starting large motors are averaged in with the power draw from all other equipment operating during the peak demand period. Spikes are typically of such short duration that they have no effect on the billing demand, regardless of the magnitude of the spike. Demand is measured in kilowatts and is the combined power of all motors and other electrical devices in use during the facility's period of greatest electrical requirement in the billing cycle. As the demand period used by most utilities is a rolling 15-minute period, the demand measured is actually integrated (added up) over the period and averaged. It represents the average sum of the power drawn in kilowatts for each motor and each electrical device that is on or running during the peak 15-minute period.

Demand is measured by the wattmeter, which integrates or sums up energy usage over the demand interval. It operates like a continuous moving average, whereby it drops off the first minute of use while adding the last minute of use during the interval. Recall from Chapter 2 that power = energy/time, and that power is measured in kilowatts and energy in kilowatt-hours (megajoules). The demand meter actually counts kilowatt-hours (megajoules) used during the interval. For example, if the wattmeter counted 550 kWh (2 000 MJ) of energy use during a 15-minute demand interval, it would report 2 200 kW of demand (550 kWh/0.25 h). Electromechanical units typically have a ratchet mechanism, whereby each new peak demand value notches the pointer up to the new demand value. Whether electromechanical or digital, the electric utility resets the demand value to zero each month after taking its reading for the monthly billing.

Although it would be difficult to determine the demand at any given time without a wattmeter, it is not hard to estimate what the maximum demand might be. It is a worthwhile exercise to estimate the WWTP's demand based on knowledge of the power draw of each piece of equipment coupled with the equipment's estimated duration of use (see Chapter 10).

The demand charge is a charge used by the utility to recover its capital or fixed costs of providing power service. The cost for repayment of debt for building power plants, transmission lines, transformers, and rights-of-way are recovered through demand charges.

The utility must size its generation capacity to serve the maximum (peak) demand on its system, plus some percentage of gross generation capacity as a safety margin and to sustain load growth. The cost of generating WWTP capacity is a fixed cost that, under utility pricing tenets, must be allocated over the customer base on a "just and reasonable" basis. The demand charge assesses each user for fixed charges based on maximum power requirements that have a direct bearing on the size of wires, transformers, and generation capacity provided by the utility for the specific service.

Like a WWTP, a generation plant is at peak efficiency at an average load factor of 100%. Consequently, it is to the utility's advantage to encourage an even load distribution over its operating schedule. In reality, a continuous 100% load factor is not possible or even a reasonable expectation. However, it is possible to build incentives into the pricing mechanism to encourage a more even distribution of energy consumption. Typically, these incentives have taken the form of pricing penalties assessed on energy consumption during periods of greatest demand. In addition, a seasonal differential is often assessed.

Billing Demand. Rates may vary for demands created during predefined peak usage periods versus off-peak or even partial peak periods. Demand charges may even change seasonally. While most demand charges are created on a billing-month basis, some utilities may include ratchet mechanisms that look at the maximum demand created over the previous 10 to 12 months and assess an *either/or* charge based on whether the current month is greater than, for example, 85% of the maximum demand from the previous 12 months.

One other aspect of demand charge is *contract demand.* This is often a minimum charge based on a contract created between the utility and a large customer, such as a WWTP, for providing initial service. This will occur, for example, when a new WWTP is built to service greater future needs. When a WWTP is built in a remote area and requires new electrical service, the electric utility may have to provide new transmission lines and transformers and size them for greater future needs. If initial usage by the facility is low, conventional demand charges will not help the utility meets its debt obligations. The utility may require that the facility owner enter into a contract through which the facility owner agrees to pay a minimum charge in the form of a higher level of demand to obtain the necessary electrical service. This additional charge helps the utility recover its capital investment at a pace more in line with its bond debt repayment.

Power Factor Adjustment. *Power factor* may be a difficult term to understand for people unfamiliar with electricity. It represents an anomaly in electrical supply. Electrical engineers refer to it as *actual* or *real* power versus apparent power, with PF being the ratio of the two. In simple terms, actual power is represented by the instantaneous power demand as measured in kilowatts; apparent power can be represented as instantaneous power as measured in kilovolt-amperes, which is the product of the voltage and current being used.

$$PF = \text{actual power/apparent power} = \text{kW/kVA} \quad (3.1)$$

Apparent power being delivered to a WWTP can exceed actual power because of the use of certain common types of devices such as transformers and inductive motors. The difference between apparent power and actual power is called reactive power (represented by the term kVAR) and is the energy used to produce the induced magnetic forces in transformers and motors. These inductive forces cause a phase shift between voltage and current. The PF is also defined as the cosine of the phase angle between voltage and current.

$$PF = \cos(\theta) \quad (3.2)$$

The problem originating from PF is that the size of transmission lines and transformers must be based on apparent power (kilovolt-amperes), but the utility can only recover the costs of real power in demand and energy charges. Consequently, the PF charge is often assessed as a penalty. This penalty, similar to demand, is used to recover the utility's capital costs for having to provide oversized equipment to satisfy customers with low overall PFs.

Utilities use various methods to assess for a low PF. It is rarely a direct charge. Often, a penalty is assessed if the PF does not meet a certain standard, for example, 85%. Another less obvious way is to assess demand charges in terms of kilovolt-amperes instead of kilowatts. If demand charge is assessed in terms of kilovolt-amperes, it is a combined charge for both demand (actual power) and a low PF.

The PF is corrected most commonly by using capacitors or synchronous motors. More detail is provided on synchronous motors in Chapter 4.

The PF is measured using a meter that is separate from the wattmeter and measures kilovolt-amperes or PF directly. The kilovolt-amperes unit is a measurement of reactive power and is represented by the third leg of the power triangle (see Figure 3.1).

MISCELLANEOUS CHARGES AND SURCHARGES. Fuel Cost Adjustment. As mentioned above, energy rate schedules are often fixed for long periods of time but adjusted monthly for the utility's fuel cost, which may be subject to frequent price fluctuations.

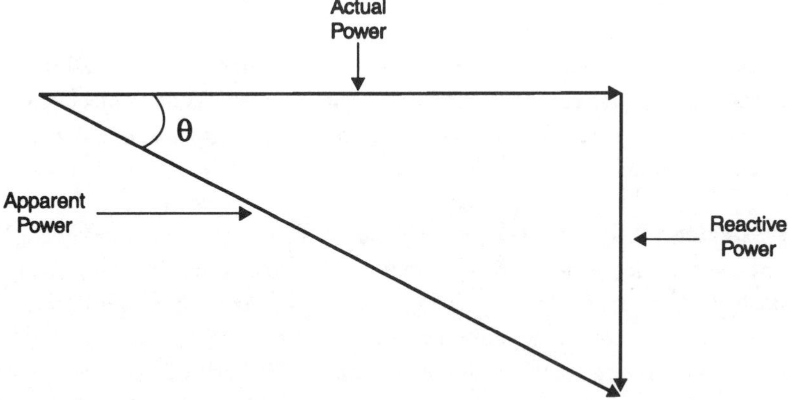

Figure 3.1 The power triangle relationship

Regulatory Fees. Regulatory fees are policy- or legislative-adjusted rates that are imposed to satisfy a cost deficiency that results from fuel cost apportionment or are mandated by legislative or other action.

State and Local Taxes. These can be applied to the consumption of both demand or energy as a government revenue source. Most municipal operations are exempt from state and local taxes.

Transmission Voltage. Discounts may be applied to demand, energy, or total amount billed when a user transforms the power to lower voltages with its own equipment and at its own expense. This recognizes the utility's lower capital cost investment. The discount may take the form of a single percentage discount or may vary with service voltage.

Standby Service. Many municipal WWTPs are directly connected to two separate sources of power for greater reliability. The second source is typically only used when the primary source fails. Standby service charges apply to the secondary source, even if not used.

Additionally, many cogeneration/independent power producers use the utility grid as a secondary source of reliability for must-run process electric loads. Some electric utilities charge for the ability to have the electric grid serve as a standby supplier of electric service in the event of forced or scheduled outage of the customer's generation equipment. These rates have the appearance of demand charges applicable to the loads that can be served by the utility.

Nonfirm Power Supply. This concept provides price incentives to users whose power supplies can be interrupted or reduced on short notice. Nonfirm rates, or interruptible rates, allow the electric utility the choice of diverting power from one customer or customer group to another.

RATE STRUCTURES. The utility has as its objective in rate design the responsibility of designing a rate that is "just and reasonable" and not unreasonably discriminatory between customers or groups in its application to the various customers served. Any rate structure that meets that objective is acceptable. In practice, there are several primary rate structures used by utilities, either singly or in combination.

Flat Demand Rate. Where demand is assumed to be known and fairly constant and metering is deemed unnecessary or not cost effective, a utility may establish a flat demand rate consisting of a price per kilowatt or horsepower (watt) over a certain duration of time. A good example of this is external lighting such as street lights, which are often billed on the basis of a flat charge per light per month, variable with the size of the light and with the estimated number of dark hours per month (by season). For example, a rate for a 100-cd (candela) (1 000-lm [lumen]) street light may be expressed as a flat charge of $50 per month, subject to a dark hours index, which adjusts the charge per month in accordance with the number of dark hours either measured or estimated by the utility or some other authoritative source.

Flat Energy Rate. The use of flat energy rates can be designed to measure direct variable costs of electric service or other production costs. The flat rate results in a level charge per kilowatt-hour of energy consumption, regardless of the consumption amount.

Straight Line Meter Rate. Small users may be assessed charges on the basis of a single price per kilowatt-hour (megajoule), regardless of total use. This kind of rate structure does not recognize the demand component of cost of service.

Block Meter Rate. The block meter rate is a common rate structure that provides variations per unit price for blocks of use that can be arranged as declining block or inverted block rate systems. The blocking structure is used to attribute lesser or greater costs to a customer, depending on the block pattern and price signal. The blocks can be tailored to affect different users differently and, therefore, may both facilitate cost recovery and motivate customers to react to price encouragement. This type of rate structure is common in the declining block form for residential and small commercial service. This can be illustrated by the sample computation in Table 3.1, assuming a total consumption of 250 kWh (900 MJ).

Table 3.1 Declining block rate structure for residential and small commercial services

Block rate	Rate, \$/kWh[a]	Computation	Amount billed
First 25 kWh	\$0.100 00	25 × 0.100 00	2.50
Next 100 kWh	\$0.060 00	100 × 0.060 00	6.00
Additional kWh	\$0.050 00	125 × 0.050 00	6.25
		For 250 kWh	\$14.75

[a] kWh × 3.600 = MJ.

EXAMPLE OF RATE WITH INCENTIVES TO LOAD SHIFT

The potential benefits of pricing incentives can be shown using the current industrial rate structure of an electric utility in northern California (Tables 3.2 and 3.3). This rate structure is an example of time- and season-differentiated pricing, which provides an incentive to electric utility customers to moderate their energy consumption during peak periods. For the service area of the utility depicted in Tables 3.2 and 3.3, the period of highest electric demand is summer weekdays (Monday through Friday, May through October), for 6 midday hours per weekday (10:00 a.m. to 4:00 p.m.). During these periods, maximum rates are assessed on both demand and energy.

Table 3.2 Industrial rate structure for an electric utility

Demand	Period A[a] demand, \$/kW-mo	Period B[b] demand, \$/kW-mo
Maximum on-peak demand	11.80	2.65
Maximum demand	2.55	2.55

[a] Period "A" = May through October.
[b] Period "B" = January through April, plus November and December.

Table 3.3 Example of time- and season-differentiated pricing

Energy	Period A hours, M-F/SSH	Period A energy, \$/kWh	Period B hours, M-F/SSH	Period B energy, \$/kWh
On-peak	6/0	0.070 44	0/0	0.000 00
Partial-peak	7/0	0.054 69	13/0	0.063 80
Off-peak	11/24	0.052 60	11/24	0.053 53

Note that no peak period has been designated during winter months (Period B) in this rate schedule.

This rate schedule provides an incentive to industrial users who have the capability of scheduling their energy consumptions to shift as much of their summer load as possible from peak periods to partial- and off-peak periods. The greatest amount of savings is possible by shifting load to off-peak periods; however, this may not always be operationally feasible.

The specific structure of seasonal and time-of-use incentives varies widely among utilities and the various states served.

SAMPLE ELECTRIC BILL. To illustrate the manner in which the above rates may be assessed, Table 3.4 is an electric bill for a pumping station during a summer month.

Table 3.4 Pumping station summer month electric bill

Billing determinant	Time-of-use	Maximum demand, kW	Maximum peak demand charge	Maximum demand charge	Total amount billed
Demand	On-peak	532	$6 278		6 278
	Partial-peak	577		$1 529	1 529
	Off-peak	497			
Total demand charges					$7 807

In this example, the local utility assesses two types of demand charges:

- Maximum peak period demand—at $10.50/kW of demand during peak hours, and
- Maximum demand during the month—at $3.10/kW of maximum demand, regardless of time of day.

For the Table 3.5 customer, energy charges were assessed on time-of-use rates on a per-kilowatt-hour (megajoule) basis, as shown. The customer charge for this class of service is $200.00 per meter, bringing total electrical costs for the sample month to the following:

Demand	7 375
Energy	14 401
Customer Charge	200
Total Amount Billed	$21 976

Table 3.5 Example of customer energy charges assessed on time-of-use rates

Billing determinant	Time-of-use	Total energy consumption, kWh	Time-of-use energy rates, $/kWh	Total amount billed
Energy	On-peak	47 008	0.070 44	3 311
	Partial-peak	81 744	0.054 69	4 471
	Off-peak	79 248	0.052 60	4 168
Total energy charges				$11 950

Utilities may apply a wide range of variations in rate structure, depending on the particular cost-of-service circumstance and service objectives (including source and cost of fuel and generation capacity requirements), the energy use objectives of the utility, and the state in which the utility resides. However, the basic components of demand, energy, and customer charge are similar throughout the U.S.

ELECTRIC SERVICE OPTIONS. The various service options available will depend on the local utility's circumstance. An example of a nonstandard power purchase arrangement offered by utilities is the interruptible service contract. As discussed previously, lower rates are sometimes offered by utilities to industrial customers that have the ability to manage their energy consumption.

- *Interruptible service* means that the customer may be required to cease all use of interruptible electrical service when requested to do so by the utility on short notice. Automatic interruptions may be required and are controlled by underfrequency relay equipment.

  A typical contract may include provisions that the utility may interrupt power deliveries to the customer no more than *x* times per month or per year, with a specified minimum notification period (for example, 30 minutes to 1 hour). This contract arrangement provides a discount to the customer in return for giving the utility more flexibility in managing its power deliveries, thus optimizing the utility's use of generation capacity.
- *Curtailable service* means that on no more than a specified number of occasions, the WWTP may be required to cut back a designated amount of electrical demand (measured in kilowatts) when requested by the utility.

COGENERATION FACILITIES

A variety of contracts may be available to customers who operate cogeneration facilities. Examples of special contracts include

- Paralleling agreement,
- Standby power agreement,
- Standard operating agreement,
- Gas supply agreement, and
- Gas transportation agreement.

Each will be described briefly below.

A paralleling agreement conveys authority to the cogenerator to operate its cogeneration unit in parallel with the utility power system. Parallel operation means operating while connected to the power grid for augmenting power requirements to greater than the generator's abilities. The interest of the utility in this contract is to protect its network and customers from disturbances that could occur from a cogenerator's output. It conveys to the cogenerator the right to operate in parallel, while allowing the utility the right to place various demands on the cogenerator.

Some utilities do not favor paralleling because they fear it may have an adverse effect on their grid systems.

A standby power agreement compensates the utility for holding a portion of its generation capacity available to serve WWTP loads, should the cogeneration unit fail or be taken off line.

A standard operating agreement spells out many of the detailed operating parameters that help maintain the integrity of the utility network. The agreement may also include rules regarding how power plant operations must maintain control and how users should pay for protection equipment that interconnects the two systems.

CONCLUSION. Previously, users have not had a choice in selecting a utility service provider, but this may be changing. It should be noted from the above that utilities use somewhat arbitrary standards for assessing users' costs. Energy charges are highest when the utility's costs are actually the lowest; demand and PF charges are averaged and are not a direct reflection of the actual capital cost required for a specific user. While the utility may have designed its rate structure to be "just and reasonable" and not unreasonably discriminatory between customers, every customer has its own individual needs. A large user such as a WWTP must take advantage of any rate break available. Consequently, differing rate structures and time-of-use charges allow for cost-saving opportunities.

NATURAL GAS

The basic rate components of demand, energy, and customer charge are similar in concept for natural gas. In addition, the same variations on rates, based on peak day and season, are generally applied.

With natural gas, energy is measured by units of gas consumed, measurable in hundreds of cubic feet (ccu ft) (cubic feet × [2.832×10^{-2}] = cubic metres), thermal heating value (therms), or millions of British thermal units (MBtu) (Btu × 1.055 = kJ).

- One therm = 100 000 Btus, and
- One MBtu = 10 therms.

Because 1 cu ft (0.03 m^3) of methane has a heating value of approximately 1 000 Btu (1 055 kJ), 1 ccu ft (3 m^3) of natural gas has a heating value of approximately 1 therm.

Recently, the trend has been to offer separate rates for the commodity itself (therms of natural gas) and for the distribution of the commodity (transportation). These changes in rate structure are in response to deregulation and to a growing number of independent gas producers that offer natural gas at discounted prices but need to move the commodity by way of existing distribution lines owned and operated by established utilities.

RATE STRUCTURES. The billing determinants for gas service vary from state to state and also depend on the industry arrangement present in the local market. The following is a general view of the diversity of industry status.

Unbundled Utility Services. Through deregulation, competitive markets have developed in some states as alternatives to the traditional monopolistic markets of the local gas suppliers. A noncore distinction is made for customers that have alternative choice(s) for procuring gas or other economic fuel alternatives. In the noncore market, gas services have been disaggregated so that gas procurement, gas transportation, delivery balancing, and storage services can be separately procured and charged.

Transportation. Some states distinguish gas services between traditional procurement–transportation–supply service and transportation service only.

Pipeline Direct. A third condition exists in some areas where FERC-regulated interstate pipelines service industrial demands with a direct connection from a transmission line and bypass the local distribution company serving the surrounding area.

The billing components of gas service and transportation service are similar to those of electric utilities as discussed above.

- *Demand* is the contracted level of service indicated by the customer's experienced peak usage on the system's peak day or a negotiated maximum until operating experience justifies a change. Demand may also be assessed using monthly peak or moving average peaks to assess the system service rendered.
- *Demand charge* is a monthly fixed charge that is applied to the measured or contract demand above.
- *Energy or throughput* is a variable rate assessed on the actual gas throughput that has been conveyed or delivered.
- *Interruptible service* in the gas industry is generally limited to large customers. Interruptions are often based on community heating load conditions. The level of interruptibility is a negotiable element in service rates.

RATES. Rate categories vary for each utility with regard to terminology and the conditions that apply to each class of service. General service rates apply to small users and are generally at a firm rate and are noncurtailable in many parts of the country. Industrial rates vary more with the season, may require a minimum usage per month year round, and vary in curtailability with the utility. Large users will find the industrial rates attractive for cost savings, particularly in summer months, but the frequency of curtailability must be tolerable by either reducing production or switching to a backup fuel.

In addition, the unbundling of natural gas rates in some states has resulted in the availability of special contracts for natural gas transportation and supply.

Gas Transportation Agreement. The gas transportation agreement covers the transportation services the utility provides in moving gas from the state border (or in-state gas fields) to the customer delivery point. The contract may provide that the customer will be served under a cogeneration transportation rate schedule, which provides that the transport cost must be no more than the equivalent service to the utility's gas boilers that generate electricity. The rate structure has components that vary with the volume of gas used and measured monthly and other components that fluctuate on a rolling basis of 12 months, measured on the highest demand at any single point in time.

Gas companies generally offer transportation programs that may be attractive to large year-round users. These are similar to commodity contracts in that the user agrees to buy a specific quantity in a specific time period at a specific price. Cost savings for a large user can be significant, but obviously the risk of committing to price, quantity, and delivery time must be carefully weighed.

The transportation of gas from well field to the local distribution company can be performed by the purchaser (in the case of large industrial loads) or by a gas broker. Once the transportation route and user's operating needs are addressed, the local gas company will assist with ultimate delivery.

The local gas company agrees to arrange transportation from the pipeline interconnect to the user typically for a fixed fee per MBtu and a fixed monthly administrative charge. The gas company may require that the customer commit to a specific quantity of a gas delivered for the following month by day 5 of each month. Gas purchased under this type of program is typically curtailable to 0%. There will generally be a minimum quantity per month to be on the program, and gas quantities used beyond the contract amount will be charged at an industrial rate.

A cost comparison example is shown in Table 3.6 for comparing a transportation program rate to an industrial rate.

SEASONAL PRICING INCENTIVE. Prices for natural gas vary with the market conditions for supply and demand. Because demand is high in winter, prices are typically highest then and lowest in the summer months. Unfortunately, WWTPs follow this pattern of high demand in winter because space heating needs and process demands such as sludge heating peak in colder months. Some WWTPs may have peak electrical power requirements in summer and may try to take advantage of low summer natural gas prices (or digester gas produced on site) to fuel engines that drive generators, pumps, or blowers. If this method is used with natural gas to *peak shave* or provide *prime* power, a more constant pattern of usage may be established over the entire year to take advantage of off-peak rates available from gas companies. A WWTP with an anaerobic digestion system that produces digester gas may need natural gas for only the coldest months; thus, it will be difficult to get anything but natural gas at the highest rates for the limited time of usage. In any case, it would be a good idea to look at the WWTP's gas usage patterns and discuss them with the gas company representative at least every 5 years to make sure that the WWTP is getting the best rate available to meets its needs.

COMPUTING THERMAL CONSUMPTION. The calculation of therms used will require a knowledge of the volume of gas used and the heating value. For natural gas, the heating value is available from the gas company and will normally range from 1 000 to 1 080 Btu/cu ft (37 000 to 40 000 kJ/m^3). Liquid petroleum gas heating values are also available from the gas supplier and are generally 92 000 Btu/gal (26×10^6 kJ/m^3).

Because gas is a compressible fluid, it is not always sufficient to know the volume of gas without knowing its pressure. Natural gas is supplied at varying pressures, and the compressed volume delivered must be converted to volume at standard conditions of temperature and pressure.

Note that the moisture content and hydrogen sulfide concentration are important in determining the corrosiveness of the gas.

For natural gas customers with an interruptible service, a backup fuel is required. For equipment such as burners for boilers, a dual fuel burner will serve this purpose. For smaller equipment such as space heaters, different

Table 3.6 Example of gas rate comparisons

Transportation program	Jan	Feb	Mar	Apr	May	Jun	Jul	Aug	Sep	Oct	Nov	Dec	Total
MCF	15 000	15 000	15 000	12 000	10 000	10 000	10 000	10 000	10 000	10 000	12 000	15 000	144 000
Administrative charge $250/mo.	$250	$250	$250	$250	$250	$250	$250	$250	$250	$250	$250	$250	$3 000
Off-peak flow, MCF	10 000	10 000	10 000	10 000	10 000	10 000	10 000	10 000	10 000	10 000	10 000	10 000	$120 000
Handling charge	$0.50	$0.50	$0.50	$0.50	$0.50	$0.50	$0.50	$0.50	$0.50	$0.50	$0.50	$0.50	
Fuel cost rate	$2.70	$2.67	$2.11	$2.04	$2.02	$1.93	$2.03	$2.13	$1.92	$2.03	$2.23	$2.57	
Transportation rate	$3.20	$3.17	$2.61	$2.54	$2.52	$2.43	$2.53	$2.63	$2.42	$2.53	$2.73	$3.07	
Transportation fuel charge	$32 000	$31 700	$26 100	$25 400	$25 200	$24 300	$25 300	$26 300	$24 200	$25 300	$27 300	$30 700	$323 800
Peak gas, MCF	5 000	5 000	5 000	2 000	0	0	0	0	0	0	2 000	5 000	24 000
Peak gas cost, $5/MCF	$25 000	$25 000	$25 000	$10 000	$0	$0	$0	$0	$0	$0	$10 000	$25 000	$120 000
Total at transportation program rates	$57 250	$56 950	$51 350	$35 650	$25 450	$24 550	$25 550	$26 550	$24 450	$25 550	$37 550	$55 950	$446 800
Industrial rate													
Handling charge	$0.53	$0.53	$0.53	$0.53	$0.53	$0.53	$0.53	$0.53	$0.53	$0.53	$0.53	$0.53	
Gas cost adjustment	$3.64	$3.64	$3.11	$3.11	$3.11	$2.97	$2.97	$2.97	$2.88	$2.88	$2.88	$2.43	
Off-peak rate	$4.17	$4.17	$3.64	$3.64	$3.64	$3.50	$3.50	$3.50	$3.41	$3.41	$3.41	$2.96	
Off-peak (off-peak flow above)	$41 700	$41 700	$36 400	$36 400	$36 400	$35 000	$35 000	$35 000	$34 100	$34 100	$34 100	$29 600	$429 500
Peak gas cost (peak flow above)	$25 000	$25 000	$25 000	$10 000	$0	$0	$0	$0	$0	$0	$10 000	$25 000	$120 000
Total at industrial rates	$66 700	$66 700	$61 400	$46 400	$36 400	$35 000	$35 000	$35 000	$34 100	$34 100	$44 100	$54 600	$549 500

burner orifices are required to fire liquid petroleum gas and different burners (and heaters) for fuel oil. Therefore, it is important to have a portion of the noninterruptible service to serve those smaller units that cannot be operated with dual fuels.

UTILITY RATE AND SERVICE OPTIONS

The best way to find out about the types of electric and gas rate structures available is to contact the local utility provider and inquire about the various service options available. Find out what kinds of tariff schedules are available for industrial customers and learn what the special requirements are to qualify for one rate schedule versus another. Obtain copies of the tariff pages and sample contracts for review and analysis.

In some cases, lower rates may be available to customers with high average loads. In other cases, special contracts may be available to customers who have some flexibility in their energy consumption patterns. For example, customers who have the ability to interrupt or curtail energy deliveries may qualify for special rates. Prepare for initial discussions with utility company representatives by studying and graphing WWTP loads—both energy and demand—for 12 consecutive months to fully understand the potential benefits of the various rate and service options offered.

Chapter 4
Electric Motors

In a typical wastewater treatment plant (WWTP) approximately 90% of the electrical energy consumed is used in the operation of motors (WPCF, 1984). Because of their high efficiency and low maintenance requirements, three-phase squirrel cage induction motors are the most common type in use. The efficiency of squirrel cage motors is affected by a number of factors, including the characteristics of the electrical power supplied and the motor's size, design, electromechanical integrity, and operating load. The last two factors can be partly controlled by the operator through proper operations and maintenance (O & M) practices. Motorized operations should incorporate the most practical energy-efficient systems available during the design process. Comparing competing systems on the basis of annual worth for capital plus O & M costs will facilitate the proper selection.

OPERATING POWER

Motors are rated in units of watts (horsepower) on the basis of the maximum amount of work that they are capable of performing. The actual power consumed at any given time by an operating motor is in direct proportion to the work being performed. In other words, if a 75 000-W (100-hp) motor is only doing 60 000 W (80 hp) of work, the motor is only supplying the equivalent electrical power of 60 000 W (80 hp). In general, motors 10 to 20% larger than actually required are usually selected by the engineer in the design process.

Nameplate power is the output power of the motor; it is also known as the *shaft* power. Input power to the motor is always greater than output power. For example, a 75 000-W (100-hp) motor doing 75 000 W (100 hp) of work will probably be drawing 80 000 W (110 hp) from the power supply. Because motors are commonly oversized by 10% and only 90% efficient, nameplate is often taken at face value for power draw in energy calculations.

SERVICE FACTOR. A 75 000-W (100-hp) motor on a pump that needs to do 110 000 W (150 hp) of work will most likely burn out. The service factor is a number placed on each motor by its manufacturer indicating the safe amount greater than the nameplate rating at which the motor can run without burning out. It is unwise to install a motor in an application with a design rating in excess of the motor nameplate. The service factor allows for above-design operating power requirements caused by such factors as bearing problems. Service factor is stamped on each motor nameplate and is expressed as a percentage of the nameplate power rating, for example 1.1 or 110%.

TYPES OF ELECTRIC MOTORS

Motors are available for direct current (dc) or alternating current (ac) applications, with ac motors also being available for single-phase or three-phase current. Motors are also available for many standard voltages. Small motors are usually single-phase and operate on 110-V electric current. The larger motors found in WWTPs are almost always three-phase and operate on the 480-V three-phase power that normally serves industrial applications.

All ac electric motors operate on the same basic principles regardless of type or size. Rotation of the shaft is the result of the force created by the interaction of a magnetic field and the current between rotor (rotating electrical coil) and stator (stationary electrical coil). It makes no difference whether the magnetic field is created in the rotor or the stator. Within this simple principle, there are many different types of ac motors, each having its own operating characteristics specifically suited to the drive application.

THREE-PHASE MOTORS. Three-phase motors are the workhorses of industry. There are three general types of three-phase motors, all having unique rotors and operating characteristics. The three types are the squirrel cage induction motor, the wound rotor induction motor, and the synchronous motor. The three-phase electric power required for these motors, regardless of whether they are induction or synchronous motors, can be calculated as follows:

$$P = V \times I \times \sqrt{3} \times \mathrm{PF} \tag{4.1}$$

Where

P	=	power, W;
V	=	line voltage, V;
I	=	average line current of 3 legs × square root of the number of phases, A (amperes); and
PF	=	power factor.

Squirrel Cage. The most common industrial motor is the squirrel cage induction motor. It is the least expensive type of induction motor and is available in all common power ratings and synchronous speeds. Voltage is applied directly to the stator or primary winding. The rotor, or secondary winding, consists of bars of aluminum or copper connected together at both ends by a conducting ring in an arrangement resembling a squirrel cage. The rotor windings form a complete closed circuit with no external connections. The absence of moving electrical contacts make these motors quite reliable.

The speed of a squirrel cage motor is nearly constant over the normal range of loads. The speed of all induction motors is determined by the line frequency, the number of poles in the motor, and the slip. The synchronous speed of the motor is the speed in synchronism with the frequency of the electric current or the speed at which the magnetic field revolves. It depends on the line frequency and the number of poles built into the motor, and is determined as follows:

$$N_s = \frac{120f}{P} \tag{4.2}$$

Where

N_s	=	synchronous speed, r/min;
f	=	frequency (60 Hz); and
P	=	number of poles per phase.

Synchronous speeds are typically available at 900, 1 200, 1 800, and 3 600 r/min. For a squirrel cage motor to produce torque, the operating speed must be less than the synchronous speed. The difference between the operating speed and the synchronous speed, which is called *slip,* is proportional to the torque produced and is usually less than 4% of synchronous speed at full load.

Squirrel cage motors are classified by the National Electrical Manufacturer's Association (NEMA) into four design types (A through D) based on torque, slip, and starting characteristics. Type A has low torque, low slip, and normal starting current. Type B is similar to type A but draws less current. Type C has more starting torque than types A and B but cannot bring all loads up to full speed. Type D has high starting torque and high full-load slip.

Wound Rotor. In the wound rotor induction motor, unlike the squirrel cage motor, the free ends of the secondary windings are brought out to slip rings through brushes and are externally connected. By varying the external resistance, the speed can be controlled. Without external resistance, this motor behaves similarly to the squirrel cage induction motor. Energy is expended in the external resistance, and efficiency varies approximately in proportion to the speed reduction. Use of the wound rotor motor is typically reserved for applications in which high starting torques and intermittent operation are required, such as for hoists and cranes.

Synchronous Motors. In a synchronous motor, dc voltage is applied to the rotor, which locks in step with the rotating magnetic field of the stator, causing the rotor to rotate at synchronous speed. However, the synchronous motor is incapable by itself of producing torque at other than synchronous speed; thus, alternative starting mechanisms, such as a squirrel cage winding, are required. In the industrial field, synchronous motors are used only in the larger power range, typically 75 000 W (100 hp) or greater.

In contrast to the induction motor, a synchronous motor's power factor (PF) is not a function of load or size but is under the control of the user. By adjusting the dc current, the PF can be changed to a leading, lagging, or unity value. Thus, synchronous motors may be used to offset the PFs of inductive loads.

SINGLE-PHASE INDUCTION MOTORS. Single-phase current is not generally produced in the U.S., but it is readily derived from any two lines of a three-phase supply system. Because of the higher costs associated with local transmission and voltage step-down of three-phase current, single-phase, low-voltage current is supplied to homes and farms. Unlike three-phase motors, single-phase motors are incapable of self-starting and require an auxiliary winding. Motors for single-phase current are more complex than three-phase motors because they require additional devices such as switches and capacitors to induce the rotating magnetic field. Consequently, single-phase motors are generally more expensive, less efficient, less reliable, and have shorter service lives than three-phase motors. For many small-power applications, however, the extra motor cost is outweighed by the costs of additional wiring and motor controls. Additionally, low-power, single-phase motors feature the added convenience of being able to be plugged in to the nearest outlet.

There are many types of single-phase motors, including shaded-coil, inductive-split-phase, capacitor, repulsion-start/induction-run, and repulsion-induction. For industrial applications, capacitor motors are typically used. The most common capacitor motors are the capacitor-start motor, the capacitor-start/capacitor-run motor, and the permanent-split-capacitor motor.

DIRECT CURRENT MOTORS. An advantage of dc motors is precision control, so they are often used for controllers in automation applications. They are also used for other applications requiring precision, such as cranes, hoists, and elevators. Additionally, they are used in battery-powered mobile equipment and for many electric railway operations. In WWTPs, large dc motors have been used in variable-speed control applications such as influent and return activated sludge pumping using rheostats or variable-voltage controllers. Because it is impractical for large power producers to generate and distribute dc voltage, voltage for dc motors is generally derived near the application from ac through various types of rectifiers.

CONVENTIONS FOR SPECIFYING MOTOR PERFORMANCE

DEFINITION OF EFFICIENCY. Motor efficiency is the ratio of motor output to motor input. It is a measure of how well a motor converts electrical energy to mechanical energy and is usually expressed as a percentage.

$$\text{Motor efficiency} = \frac{P_o}{P_i} \times 100 \tag{4.3}$$

and

$$P_o = P_i - P_L \tag{4.4}$$

Where

P_o = mechanical power output, W;
P_i = electrical power input, W; and
P_L = power loss, W.

Power loss is that part of the input power that is converted to heat rather than useful work. This loss can be attributed to friction and windage, electrical resistance losses in the rotor and stator, magnetic core losses (hysteresis and eddy currents), and stray load losses.

TEST PROCEDURES. The Institute of Electrical and Electronics Engineers (IEEE) Standard IEEE-112 (IEEE, 1984) is the recognized test procedure in the U.S. This standard is not a single test, but specifies five different methods—A, B, C, E, and F (there is no D)—each of which is applicable to a specified type and size of motor.

The NEMA has recommended IEEE-112, Method B, for motors ranging from 750 to 90 000 W (1 to 125 hp). In this method, mechanical output is measured by loading the motor with a dynamometer. Method A requires that a mechanical brake be used to load the motor. Because of heat buildup, this method is typically reserved for fractional power motors. Method C involves coupling two identical motors together so that one acts as a generator. Losses are determined by measuring the input and output of the motor/generator setup. This method is typically restricted to motors of 225 000 W (300 hp) or greater and usually performed only by customer specification. Methods E and F are generally used for large motors. These methods do not require that output be measured. Instead, efficiency is determined by calculating the individual losses.

Caution should be exercised when comparing efficiencies based on U.S. standards with those of foreign manufacturers. The two most common test standards employed by foreign manufacturers are

- International Electrotechnical Commission, *Rotating Electrical Machines*, IEC 34-2; and
- Standard of the Japanese Electrotechnical Committee, *Induction Machines,* JEC-37.

Application of these standards results in efficiencies that are typically 0.5 to 2% higher than ratings using IEEE Standard 112; thus, they are not directly comparable (Andreas, 1982; Cummings, 1982; Cummings *et al.*, 1981; and Jordon, 1983).

MOTOR RECORDS. Detailed specifications for all motors in a WWTP should be on file with the maintenance records. Motor specifications found on the nameplate usually include the following: manufacturer's name, model number, serial number, rated power, voltage, number of phases, full-load amperage, speed, and service rating. The manufacturer of the motor should provide separately the motor efficiency and PF at various percentages of the full-load rating, together with recommended preventive maintenance requirements and replacement part catalog numbers.

Recommended energy conservation measures (ECMs) are as follows:

- If requirement data are not on file, they should be obtained from the manufacturer.

- Before a motor is placed into service, the nameplate information must be recorded or checked against the design specifications. Motor nameplates become illegible with time because of paint, scraping to remove paint, oil, weather, submersion in wet wells, and other factors. Therefore, the only definite way to have the information when the motor must be serviced or replaced is to write it down when it is installed.

MOTOR FAILURE. Theoretically, a squirrel cage electric motor should never fail because the only parts that wear are the wearing surfaces, and a properly loaded modern bearing will run practically indefinitely, provided it receives proper care. The magnetic lines of flux that cause the motor to operate cause no wear; the electricity flowing through the wire does not wear out the wire. Therefore, the motor should never wear out and fail. This theory applies only to squirrel cage motors because the brushes and slip rings in the other types are subjected to wear, which will eventually cause motor failure.

The major cause of motor failure is neglect, including failure to keep the motor vents clean, which results in overheating; improper lubrication (either too much or too little), which causes bearing failure and may result in complete destruction of the motor if the rotor comes in contact with the stator; and improper belt tension, which results in bearing failure and possible destruction of the motor. Failure to maintain proper overload relays and other protective equipment may cause an electrical overload. Heat and chemical action from excess oil cause breakdown of the insulation and shorted windings. Failure to maintain the driven machinery results in overloading, overheating, and, finally, motor failure. Lack of proper protection against the environment results in motor failure from damage by rust, corrosion, or contamination from foreign particles. Surprisingly, many motors are damaged by water during washdown by careless WWTP personnel. Motor control circuits are often neglected until a failure occurs. Loose connections in wiring circuits cause overcurrent or phase failure, and overheating. These are a few, but not all, of the forms of neglect that result in motor failure (WPCF, 1984). The design life of good bearings should be at least 100 000 hours (11.4 years of continuous duty). The major cause of bearing failure is excessive lubrication.

Recommended ECMs are the following:

- Periodic lubrication by trained technicians is preferable to lifetime lubrication.
- All motor bearing casings should be fitted with a grease relief to permit exit of excess grease. If the grease relief is not open, the excess grease will cause high bearing temperatures and premature failure. Excess grease may also be forced inside the motor and coat the windings, causing them to overheat.
- The following items should be checked periodically to ensure long bearing life: external or internal grease accumulations, dirt buildup,

excessive motor temperature (especially bearing ends), shaft alignment, bearing play (run-out), and excessive noise or vibration. Frequently checking the mechanical operation of a motor and driven equipment for vibration, noise, high temperature, or other abnormal conditions and providing proper routine maintenance of dust removal and bearing lubrication are extremely important for ensuring long motor life.

- The electrical operation of the motor affects the cost of operation and the ability of the motor to provide dependable service. The motor should be checked for proper voltage, amperage, and resistance to ground. How these measurements are made will be covered in detail in the following section. The operation of the motor controls should also be checked carefully. Motor starters should engage fully and not chatter. Weak coils and worn contacts should be replaced. Once per year, all connections in the control panel should be checked for tightness.

MEASUREMENT OF ELECTRICAL CHARACTERISTICS. The characteristics of the power supplied to a motor and the behavior of the electricity within the motor can be measured. These measurements are used to assess the quality of the power supply, the load being placed on the motor, the motor's general operating efficiency, and the motor's electrical integrity.

Voltage. For diagnostic purposes, three-phase voltage measurements are taken for the three leads (L_1, L_2, and L_3) on the load side of the motor controls (Figure 4.1). The voltage unbalance for the measurements obtained between L_1 and L_2, L_1 and L_3, and L_2 and L_3 should be less than 1%.

$$\text{Unbalance (\%)} = \frac{\text{Maximum deviation average}}{\text{Average of the three readings}} \times 100 \qquad (4.5)$$

If the unbalance exceeds 1%, the amperage unbalance and the motor operating temperature will be high. Frequently, problems with the motor controls or power supply result in voltage unbalances. Testing for voltage unbalance should be part of equipment preventive maintenance.

Motors are designed to operate within a narrow voltage range. Typically, the acceptable voltage range for a motor is the nameplate rating ±10%. The manufacturer's literature for a specific motor should be consulted to ensure proper operation.

Amperage. The currents (amperage) flowing through each phase, L_1, L_2, and L_3, of a three-phase motor must be close to equal. When the amperage between the phases is unbalanced, the motor runs hot and loses efficiency. Amperage measurements should be taken when the motor is at its normal operating temperature and loading. The percent unbalance is calculated using the same formula used for voltage unbalance. At full load, the maximum

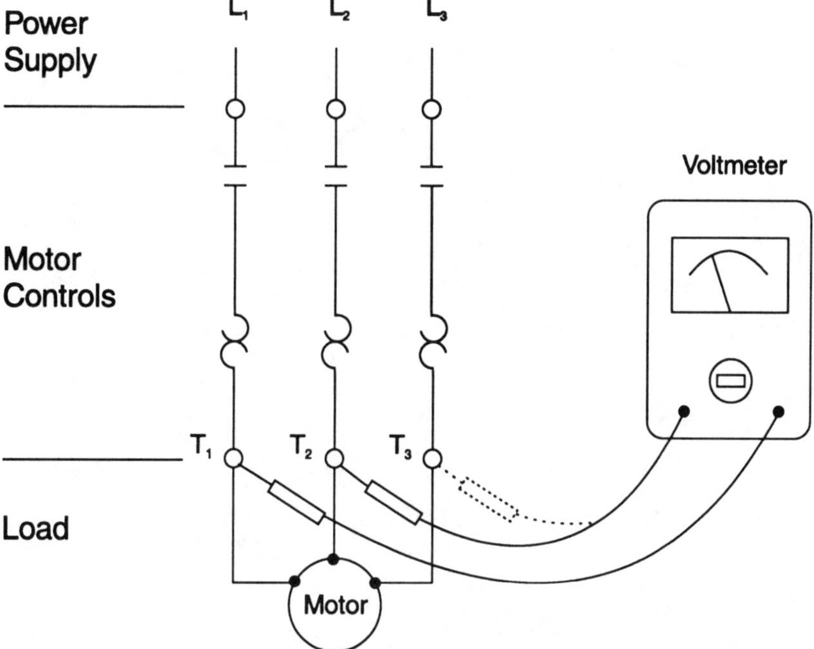

Figure 4.1 **Checking three-phase motor voltage (L = power lead and T = motor starter terminal) (reprinted with permission from ITT FLYGT Corp. [1985] *FLYGT Product Education Manual.* 8th Ed., Product Education, Norwalk, Conn.)**

allowable unbalance is 5%. If a motor is operating at considerably less than full load, a larger percentage of unbalance may be acceptable.

Identifying the source of voltage or amperage unbalance requires some troubleshooting. First, note the voltage and amperage with the original arrangement of power leads. Next, change the location of all three power leads on the load side of the motor controls and repeat the measurements. It is important to make sure that the positions of all three leads are changed. If only two leads are changed, the motor will operate in the reverse direction. Change all three leads to produce the third possible arrangement and take a third set of readings. Figure 4.2 is an example in which the black lead always has an amperage of 19. This indicates that the problem is in the motor or the wiring leading to it. If L_1 always produces the high reading, the problem is somewhere between that terminal and the power plant. The problem could be as simple as worn starter contacts or a loose terminal screw. Conversely, assistance may be required from the power company for a solution.

Power Factor. The easiest and most accurate method available to measure the PF is to use a PF meter. The PF is the ratio of the actual power (kW) to the apparent power (kVA). This relationship is illustrated by the power triangle

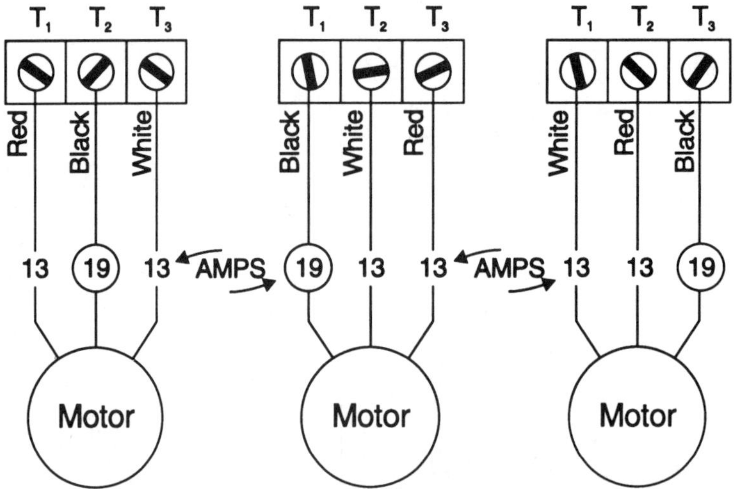

Figure 4.2 **Unbalance cause by load (T = motor starter terminal) (reprinted with permission from ITT FLYGT Corp. [1985]** ***FLYGT Product Education Manual.*** **8th Ed., Product Education, Norwalk, Conn.)**

(see Figure 3.1). The reactive power (kVAR) is the energy used to produce the induced magnetic forces that make the motor rotate. The reactive power is not measurable with a wattmeter. The PF can also be obtained by calculation, using data obtained with a wattmeter, voltmeter, and ammeter. The real power (kW) and apparent power (kVA) are measured directly and used in the calculation.

Resistance and Insulation. The flow of electricity from a point of high potential to a point of low potential is governed by the resistance of the space between the two points. An electrical conductor provides a path of low resistance to facilitate the flow of electricity along an intended path with minimal losses. Insulation confines the flow of electricity within its intended path by shielding the electrical path along a conductor with a barrier of high resistance. Over time, the insulation around conductors begins to break down and allow current losses to ground. The insulation on the wires leading to the motor and on the motor coils themselves should be checked routinely as part of preventive maintenance. This simple measurement using a megohmmeter ("megger") can help prevent emergency maintenance situations. Because the resistance of the insulation to the flow of electrons is being measured, the resistance values are measured in megohm (million ohm) units. Whenever resistance is being measured, all sources of power must be removed from the circuit being tested. Always check for voltage before putting the megohmmeter probes on the terminals. The megohmmeter has its own power supply to energize the circuit being tested. Each power lead to the motor is checked as shown in Figure 4.3. A new motor should read infinite resistance. Over time, the resistance will gradually decrease. This information can be graphed so that the trend can be visualized. When a minimum resistance value is reached, the coils can be dipped and rebaked to restore the insulation value. Some manufacturers allow 2 MΩ/hp as the minimum resistance value before the coils are serviced. For small motors, a somewhat higher ratio should be used. If the motor is checked at the motor controls, this activity is also checking the insulation of the conductor to the motor. Before the motor coils are faulted for low resistance, additional readings must be made at the motor.

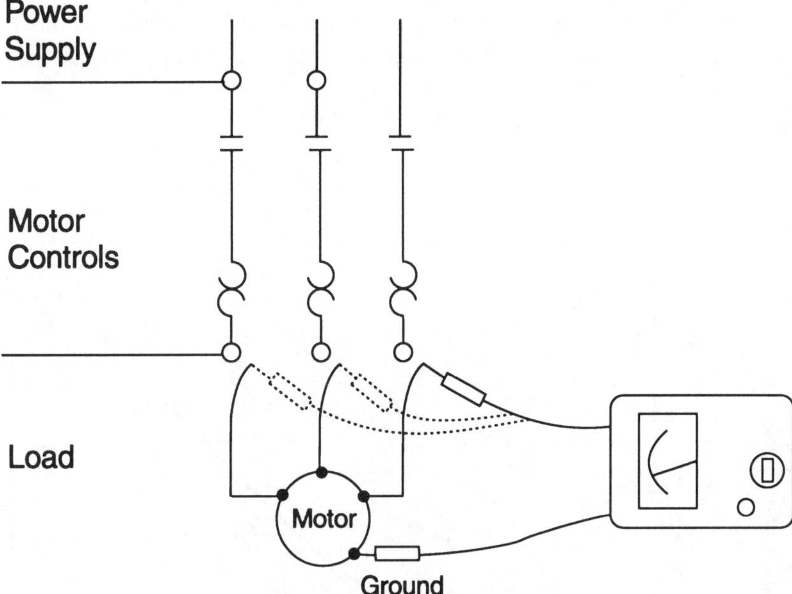

Figure 4.3 Checking motor insulation, coil to ground, for a megohmmeter (reprinted with permission from ITT FLYGT Corp. [1985] *FLYGT Product Education Manual.* 8th Ed., Product Education, Norwalk, Conn.)

Power. Measured power is the actual instantaneous electrical demand required to operate a motor under a full or partial load condition. Power is measured in kilowatts and may be obtained directly with a wattmeter. Knowing the power consumed by a motor can be useful in evaluating the operating efficiency of the system. Power measurements can be compared with the motor's power rating to determine whether the motor is overloaded (undersized) or underloaded (oversized). Overloading may be the result of poor motor selection or of a motor, drive, or driven element problem, which requires immediate attention. Motors are most efficient at their full-load condition.

Slip. Load estimates can be made from motor operating characteristics such as current, slip, and input power. Figure 4.4 shows typical performance curves for a 7 500-W (10-hp) squirrel cage motor. As can be seen, current is directly proportional to output power in only a narrow range—approximately 70 to 110% of full load. Thus, overestimates of load will be made at low loads. Slip, however, is directly proportional to load throughout the operating range and allows a method of load determination at any operating point. Load can also be calculated from input power, but the efficiency at the operating point must be known. Load determination from slip and input power measurements are discussed below.

Slip measurements provide a convenient means of estimating motor loading, provided accurate and precise measurement devices are available. Slip is the difference in actual rotor speed and the synchronous speed of the motor,

$$\text{Slip} = N_s - N \tag{4.6}$$

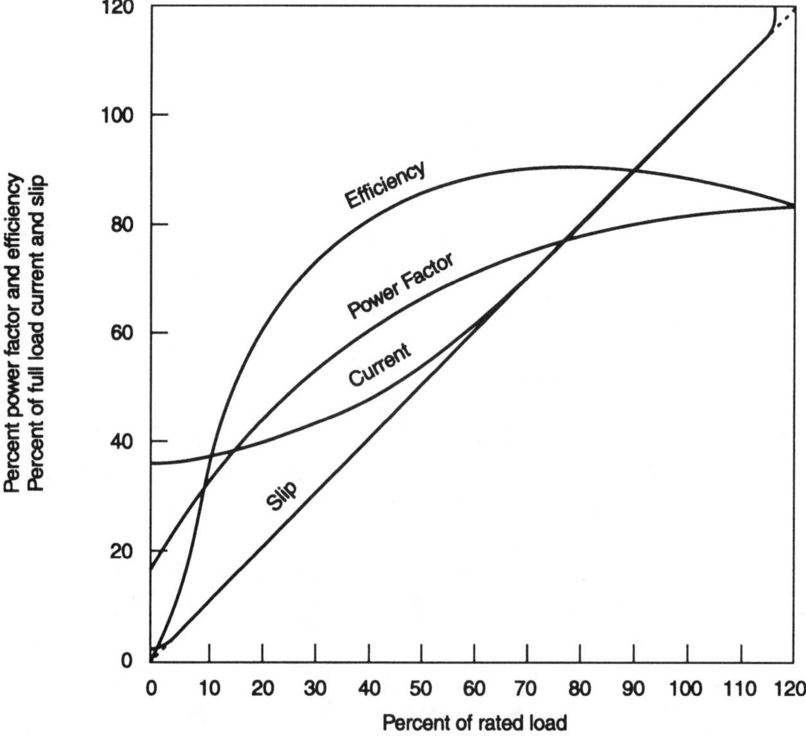

Figure 4.4 **Typical performance characteristics for a 10-hp (7 500-W) squirrel cage induction motor as a function of load**

Where

N_s	=	synchronous speed, r/min; and
N	=	actual speed, r/min.

Because slip is proportional to load, the ratio of the measured slip to the full-load nameplate slip yields the relative motor load. Slip also varies inversely with the square of the applied voltage, necessitating a correction. The load can be determined as follows:

$$P_{\text{out}} = \frac{N_s - N}{N_s - N_R}\left[\frac{V}{V_R}\right]^2 P_R \tag{4.7}$$

Where

P_{out}	=	output power, hp (W);
P_R	=	nameplate-rated full-load output power, hp (W);
N_R	=	nameplate full-load speed, r/min,
V_R	=	nameplate-rated voltage, V; and
V	=	actual voltage, V.

The slip of a squirrel cage induction motor is small (usually less than 3%). A stroboscopic tachometer with accuracy of 0.1% or better is required to produce meaningful results. An error of only 1% would result in an error of 30% in the slip measurement. Furthermore, considerable error may be inherent in this method when using manufacturers' nameplate data because NEMA standards allow a 20% deviation in the nameplate slip.

MATCHING MOTORS TO LOAD

The primary goal in selecting a motor is to be sure the motor selected will have adequate power to drive the intended load throughout the entire anticipated load range. A secondary goal is that the motor be energy efficient and economical.

Motor efficiency for standard induction-type motors increases with increasing motor size and is relatively constant for a given motor at various loads. However, most motors are slightly more efficient at full load, with efficiency dropping off slightly as load decreases from full to half load. For motors of greater than 750 W (1 hp), the change in efficiency from full to half load is generally less than 5% of the full-load efficiency. Similarly, a PF for a standard induction motor increases with motor size, but the PF drops off more significantly with a reduction in load. The decrease in PF for a 750-W (1-hp) motor at half load is approximately 25% of the PF at full load. Table 4.1

Table 4.1 Typical operating characteristics for premium-efficiency totally enclosed, explosionproof, corrosion-duty motors (WPCF, 1984)

	Speed rpm		% Efficiency			% Power factor			Current in amperes[a] 460 V			Torque at full voltage			
												Locked (starting)	Pull out (breakdown)		
hp	Syn.	Full load	Full load	3/4 load	1/2 load	Full load	3/4 load	1/2 load	Full load	Locked (starting)	Full load torque at full load speed, lb-ft	Percent of	Full load	Max kvar[b]	Code
1	1 800	1 740	85.5	85.0	82.5	78.0	70.5	58.0	1.50	11.5	3.0	275	300	0.4	L
	1 200	1 150	81.5	81.0	77.0	62.0	52.5	40.2	1.85	10.5	4.5	170	265	0.8	K
	3 600	3 500	83.5	83.0	79.5	86.5	81.5	71.5	1.95	18.5	2.2	175	250	0.3	L
1-1/2	1 800	1 740	86.0	86.0	83.5	77.0	69.0	56.0	2.25	18.5	4.5	250	280	0.6	L
	1 200	1 170	85.5	85.0	82.5	67.5	58.5	45.5	2.45	15.5	6.7	165	250	1.0	K
	3 600	3 490	84.5	85.0	82.5	89.0	85.5	77.0	2.5	24.0	3.0	170	240	0.3	L
2	1 800	1 730	86.0	86.0	84.5	78.0	70.5	57.0	2.90	23.5	6.0	235	270	0.8	L
	1 200	1 170	87.0	96.5	84.5	67.5	59.0	46.0	3.2	21.5	8.9	160	240	1.3	K
	3 600	3 515	87.0	87.0	85.0	84.5	79.5	70.0	3.8	30.5	4.4	160	230	0.7	K
3	1 800	1 755	88.5	88.5	87.0	78.0	71.5	59.0	4.15	31.0	8.9	215	250	1.1	K
	1 200	1 170	89.0	88.0	87.0	69.5	61.5	49.0	4.5	29.0	13.4	155	230	1.7	J
	3 600	3 495	87.5	78.5	88.0	88.0	84.5	77.0	6.1	45.5	7.5	150	215	0.9	J
5	1 800	1 745	88.5	90.0	89.5	83.0	78.0	67.5	6.5	45.0	15.0	185	225	1.4	J
	1 200	1 165	89.5	90.0	88.5	72.5	66.0	54.0	7.2	43.5	22.5	150	215	2.5	H
	3 600	3 495	89.0	89.5	88.5	85.0	81.5	73.0	9.3	61.0	11.2	140	200	1.8	H
7-1/2	1 800	1 760	91.0	91.5	90.5	83.0	79.0	70.0	9.5	62.5	22.3	175	215	2.0	H
	1 200	1 175	90.5	91.0	90.0	82.0	77.5	68.0	9.5	63.5	33.5	150	205	2.2	H
	3 600	3 495	89.5	91.0	90.5	88.5	86.5	81.5	11.8	79.4	15.0	135	200	1.6	H
10	1 800	1 755	91.0	92.0	92.0	85.0	81.5	74.5	12.4	77.0	29.8	165	200	2.2	H
	1 200	1 170	90.2	91.0	91.0	84.0	81.0	73.5	12.3	81.0	44.8	150	200	2.5	H

Table 4.1 Typical operating characteristics for premium-efficiency totally enclosed, explosionproof, corrosion-duty motors (WPCF, 1984) (continued)

	Speed rpm		% Efficiency			% Power factor			Current in amperes[a] 460 V			Torque at full voltage			
												Locked (starting)	Pull out (breakdown)		
hp	Syn.	Full load	Full load	3/4 load	1/2 load	Full load	3/4 load	1/2 load	Full load	Locked (starting)	Full load torque at full load speed, lb-ft	Percent of	Full load	Max kvar[b]	Code
	3 600	3 530	90.0	90.5	89.5	88.0	86.5	81.5	17.7	115.5	22.3	130	200	2.5	G
15	1 800	1 770	92.5	93.0	92.5	83.0	79.5	71.0	18.8	111.5	44.4	160	200	4.0	G
	1 200	1 175	91.0	91.5	91.0	81.0	76.5	67.0	19.0	113.0	66.7	140	200	4.8	G
	3 600	3 515	90.5	91.5	91.0	89.5	89.0	85.5	23.1	144.0	29.8	130	200	2.7	G
20	1 800	1 770	93.0	94.0	93.5	85.5	83.5	78.0	24.0	140.0	59.3	150	200	4.5	G
	1 200	1 175	91.0	92.0	92.0	83.0	80.0	72.0	24.7	140.5	89.2	135	200	5.5	G
	3 600	3 535	90.5	91.5	90.5	88.0	87.0	82.5	29.4	167.0	37.1	130	200	4.4	G
25	1 800	1 775	93.0	93.5	93.5	85.5	83.0	76.0	30.2	171.5	74.0	150	200	5.5	G
	1 200	1 185	92.5	93.5	93.5	80.5	78.0	70.0	31.3	173.5	110.8	135	200	8.0	G

[a] Full load current is listed for 460 V; 230 V-A = 2 × 460 A.
[b] For power factor correction to the maximum allowable limit—95%.
NOTE: hp × 0.745 7 = kW; lb-ft × 1.356 = N·m.

shows efficiencies and PFs at full load, three-quarter load, and half load for standard induction motors.

Power factors decrease 10 to 15 percentage points or more as load is reduced from 100% to 50%, depending on the power and revolutions per minute of a given motor. Because electrical utilities often assess charges for a poor PF, it is advantageous to closely match motor size to load to maintain a high PF at a WWTP. The relationship of both efficiency and PF to changes in motor load differ between the motor enclosure (open drip-proof, totally enclosed fan-cooled, and others), number of revolutions per minute, and manufacturer.

When studying equipment, if a significant mismatch in the load and electric motor rating is found, it may be economical to replace the motor in question with a motor that more closely approximates the required load. For example, a 15 000-W (20-hp), 1 200-r/min, high-efficiency motor is driving a pump and found to be operating at approximately half load. Pump tests verify that a 7 500-W (10-hp) motor would be able to operate throughout the operating range of the pump as installed without being overloaded. It is assumed that the motor operates at a constant load of 7 500 W (10 hp). An estimate of savings would be obtained by replacing the 15 000-W (20-hp) motor with a 7 500-W (10-hp) motor of the same type. From Table 4.1, a 15 000-W (20-hp) motor operating at half load is estimated to have an efficiency of 92%. A 7 500-W (10-hp) motor driving the same pump would be at full load; therefore, it would have an efficiency of 90.2%. This example illustrates that providing a better match of motor size to the load could result in an increase in energy consumption and corresponding increases in power costs. Those calculations are shown below.

Energy cost for existing conditions, 20-hp motor with 10-hp load:

$$\text{Power demand} = \frac{10 \text{ hp} \times 0.746 \text{ kW/hp}}{0.92 \text{ motor efficiency}} = 8.11 \text{ kW}$$

$$\text{Energy cost} = 8.11 \text{ kW} \times 8\,760 \text{ hr/yr} \times \$0.08/\text{kWh} = \$5\,683/\text{yr}$$

Energy costs for 10-hp motor with 10-hp load:

$$\text{Power demand} = \frac{10 \text{ hp} \times 0.746 \text{ kW/hp}}{0.902 \text{ motor efficiency}} = 8.27 \text{ kW}$$

$$\text{Energy cost} = 8.27 \text{ kW} \times 8\,760 \text{ hr/yr} \times \$0.08/\text{kWh} = \$5\,796/\text{yr}$$

Increased energy cost using the 10-hp motor would be $103 per year.

Again referring to Table 4.1, it is readily seen that the PF is a different story. The 7 500-W (10-hp) motor at full load would have a PF of 0.84, whereas the 15 000-W (20-hp) motor at half load would have a PF of 0.72. Therefore, a 17% improvement in PF would result from using the 7 500-W

(10-hp) motor for the 7 500-W (10-hp) load. For WWTPs located in an area where the power company or utility imposes a charge for a poor PF, this difference would have a financial effect on the WWTP.

REPLACEMENT WITH ENERGY-EFFICIENT MOTORS

Energy-efficient motors differ from normal-efficiency motors because they use electrical energy more efficiently to generate shaft output power. At present, there are no standards that define an energy-efficient motor. Manufacturers are free to apply the term to virtually any motor. In general, manufacturers use the term *energy efficient* to designate motors that have higher efficiencies than standard motors of otherwise equivalent ratings.

Initially, the typical difference between an energy-efficient and a standard motor may not seem significant. It must be understood, however, that even a difference of 1 or 2% can repay the extra cost of an energy-efficient motor in a few months or years. The more hours per day a motor runs, the quicker the repayment is and the greater the savings will be.

Two situations must be considered when evaluating the use of energy-efficient motors: the initial purchase price of an energy-efficient motor versus the cost of a standard-efficiency motor, and the changeout of an operating standard-efficiency motor. The economics for these two conditions are quite different.

It is not practical or possible to develop a rule-of-thumb to estimate the capital cost difference between an energy-efficient motor and a normal-efficiency motor. It is always necessary to check with the motor suppliers and obtain prices for high-efficiency and normal-efficiency motors for any given application. The price of a motor is a function not only of its power output, but also of its efficiency, the service factor, number of revolutions per minute, enclosure, and motor design (that is, normal starting torque, high starting torque, high slip, and others). There are so many variables that it is difficult to establish an overall rule of when one would be more economical than the other. Despite that caution, it can be said that the cost of energy-efficient motors usually ranges from 15 to 60% more than that of standard motors. It is important, here, to note that the difference in efficiencies between high and standard decreases with increasing motor size. The increase in efficiency between a high-efficiency, 750-W (1-hp) motor at full load over a standard one is approximately 12%, whereas the difference between 75 000-W (100-hp) motors is only 3%. Another important point in considering an energy-efficient motor is the proposed running time for the piece of equipment in question. The more hours each day that the motor runs, the more rapid the payback will be from a high-efficiency motor. Candidates for new high-efficiency motors would generally include any motors that run most of the day.

There are times when it is not practical to specify a high-efficiency motor. A standard low-priced motor–pump combination such as a sump pump often is not a good candidate for a high-efficiency motor. A sump pump with a submersible motor has special needs, and the concerns of a submersible, leak-proof installation or special cooling requirements may take precedence over high efficiency. Additionally, two-speed motors may have high cost premiums when ordered in high-efficiency configurations. It is always advisable to check with equipment manufacturers before specifying high-efficiency motors for special applications.

In addition to higher efficiencies obtained by using high-efficiency motors, it is generally found that high-efficiency motors have better PFs than standard-efficiency motors. This can also generate cost savings, depending on the power cost structure.

A word of caution should be given regarding the computation of squirrel cage induction motor efficiencies. The several methods available for measuring electric motor efficiencies are not consistent from method to method. In the U.S., motors are rated using a standard NEMA test, MG1-12.53A. The motor should be tested with IEEE standard 112 test method B with segregated loss determinations. It is possible that some manufacturers, especially foreign manufacturers, will employ different test methods. These different methods may yield motor efficiency ratings that are not directly comparable with U.S. ratings. Furthermore, it is important to distinguish between nominal efficiency and guaranteed minimum efficiency ratings. To calculate cost savings, the same efficiency rating should be used when comparing two motors. It is recommended that nominal efficiency ratings be used to compare standard- and high-efficiency motors for a given application. The guaranteed efficiency should be used to call out the minimum efficiency accepted for a high-efficiency motor when developing purchasing specifications. The most important thing to remember when comparing two motors for a given application is to ensure that the same standard (whether nominal or guaranteed minimum) is used and that standards are comparable only when the efficiencies are developed using the same testing method.

For example, a standard 7 500-W (10-hp) motor is to be used to drive a pump. The motor operates continuously at full load. The pump and motor are in a dry well; therefore, an open drip-proof motor can be used. Should a standard- or a premium-efficiency motor be used for this pump? The calculations to compare the two are shown below.

Motor costs:

10-hp, open, drip-proof, energy-efficient motor
1 170 rpm, efficiency = 91.2% (nominal), cost $781

10-hp, open, drip-proof, normal-efficiency motor
1 170 rpm, efficiency = 86.5% (nominal), cost $518

Cost premium for energy-efficient motor = $263

Power costs:

Power costs for continuous, full-load operation, at $0.08/kWh:

$$\text{High-efficiency power demand } \frac{10 \text{ hp} \times 0.746 \text{ kW/hp}}{0.912} = 8.18 \text{ kW}$$

Annual power cost 8.18 kW × 8 760 hr/yr × $0.08/kWh = $5 733/yr

$$\text{Normal-efficiency power demand } \frac{10 \text{ hp} \times 0.746 \text{ kW/hp}}{0.865} = 8.62 \text{ kW}$$

Annual power cost 8.62 kW × 8 760 hr/yr × $0.08/kWh = $6 041/yr

Annual power cost savings for high-efficiency motor:

$6 041/yr – $5 733/yr = $308/yr

Therefore, the payback for the additional cost to purchase a high-efficiency motor is

$$\frac{\$263 \text{ additional cost}}{\$308 \text{ savings per year}}$$

Payback = 0.85 years or 10.2 months.

The savings achieved by using a high-efficiency motor versus a normal-efficiency motor for this initial installation would be $308 per year. This compares with the initial cost premium for a 7 500-W (10-hp) motor of approximately $263. In this case, the payback would be 0.85 years, or 10.2 months, for continuous operation.

Now consider the economics of substituting an energy-efficient 7 500-W (10-hp) motor for an existing 7 500-W (10-hp) normal-efficiency motor after it has been installed and running for some time. Because the existing normal-efficiency motor is assumed to be running satisfactorily, both the cost of purchasing a new 7 500-W (10-hp) high-efficiency motor and the cost of the existing motor must be considered. For this example, no salvage value is taken for the standard motor. Therefore, the annual savings remain the same at $308 per year. The cost, however, is $781 for the new motor. Here the payback would be as follows:

$781 additional cost for new motor
$308 savings per year
Payback = 2.33 years; as an investment, this would represent approximately a 43% simple interest return on the investment.

It should be emphasized that the cost and savings data must be calculated for each specific motor considered for replacement. Because the motor cost depends on the enclosure as well as the revolutions per minute, the voltage, service factor, and other factors that affect the cost of an individual motor must be considered. The efficiencies for each of the motor types and each possible variation in the number of revolutions per minute will be different.

References

Andreas, J.C. (1982) *Energy Efficient Electric Motors, Selection and Application.* Marcel Dekker Inc., New York, N.Y.

Cummings, P.G. (1982) Comparison of IEC and NEMA/IEEE Motor Standards, Part I. *Inst. Electr. and Electr. Eng. Trans. Ind. Appl.*, **IA-18,** 471–478.

Cummings, P.G., *et al.* (1981) Induction Motor Efficiency Test Methods. *Inst. Electr. and Electr. Eng. Trans. Ind. Appl.*, **17,** 253–272.

Institute of Electrical and Electronics Engineers (1984) *IEEE Standard Test Procedure for Polyphase Induction Motors and Generators.* Std 112-1984, New York, N.Y.

Jordon, H.E. (1983) *Energy Efficient Electric Motors and Their Application.* Van Nostrand Reinhold Co. Inc., New York, N.Y.

Water Pollution Control Federation (1984) *Prime Movers: Engines, Motors, Turbines, Pumps, Blowers & Generators.* Manual of Practice No. OM-5, Washington, D.C.

Suggested Readings

Cochran, P.L. (1989) *Polyphase Induction Motors, Analysis, Design, and Application.* Marcel Decker, Inc., New York, N.Y.

Croft, T.C., and Summers, W.I. (Eds.) (1987) *American Electricians' Handbook.* McGraw-Hill, Inc., New York, N.Y.

Lobodovsky, K.K., *et al.* (1989) Field Measurements and Determination of Electric Motor Efficiency. *Energy Eng.*, **86,** 3, 41–53.

Nailen, R.L. (1989) Can Field Tests Prove Motor Efficiency? *Inst. Electr. Electr. Eng. Trans. Ind. Appl.*, **25,** 391–396.

National Electrical Manufacturer's Association (1987) *NEMA Standard for Motors and Generators.* Publ. No. MG1-87, Washington, D.C.

Owen, W.F. (1982) *Energy in Wastewater Treatment.* Prentice-Hall, Inc., Englewood Cliffs, N.J.

University of Florida (1986) *Operations and Training Manual on Energy Efficiency in Water and Wastewater Treatment Plants, Volume 1, Energy Conservation Measures.* Training, Res. Educ. Environ. Occupations Cent., Gainesville, Fla.

University of Florida (1988) *A Practical Guide to Energy Conservation Measures in Water Quality Systems.* Training, Res. Educ. Environ. Occupations Cent., Gainesville, Fla.

Chapter 5
Pumps

While pumps account for major energy uses in wastewater treatment systems, most features of pump system efficiency are determined during design and construction. There are, however, a number of operation and maintenance practices that can be implemented to either improve pumping system efficiency or restore original system efficiency. It is impossible to simply look at a pump and determine whether it is performing in an energy-efficient manner. Inefficient pumping operations do not generally smell bad, sound alarms, give off smoke, or fill customers' basements with wastewater—they simply waste energy and money. Calculations, field measurements, analyses, and research data recorded in pump curves are sometimes required to determine whether a pump is operating efficiently and evaluate what can be done to conserve energy. Because considerable effort is required to locate the source of pump inefficiency, improving pumping efficiency is often a neglected issue at wastewater treatment plants (WWTPs).

OPERATION AND MAINTENANCE PRACTICES

Pumps require routine inspection to ensure that neither packings nor seals are leaking excessively, that motors are not overheating, that excessive vibration is not occurring, and that bearings are properly lubricated. Additionally, pumps should be inspected routinely for impeller wear and pitting, casing wear, ring wear, and misalignment. With proper care, alignment, and repair or replacement of worn and pitted parts, pumps will perform at or close to their design ratings. With well-maintained pumps, energy efficiency is strictly a matter of system design, system condition, and how the pumps are operated. In some simple systems, little can be done operationally to affect pumping efficiency, while in complex systems, several things can be done to improve efficiency. Most systems fall in between. The following discussion forms the basis for understanding how and where operational strategies can come into play to positively affect pump system efficiency.

Energy conservation measures for pumps are as follows:

- Proper maintenance will keep the pump at or near its original design efficiency rating.
- Proper setting of controls can enable pumps to work more efficiently (Pincince, 1970; Yin *et al.*, 1996).

PUMPING PRINCIPLES

Pumps are used in wastewater collection systems to raise water to high points in the system, to which the water cannot flow by gravity. Even where gravity flow is possible, it is sometimes more economical to pump water through a force main because a force main can be considerably smaller than a gravity main of the same capacity. In addition, pumps are used to raise wastewater from sewers that flow into most WWTPs, enabling it to flow by gravity through the plant.

Several references are available on wastewater pumps (ASCE, 1992; Hicks and Edwards, 1971; Hydraulic Institute, 1983; Karassik *et al.*, 1976; Sanks, 1989; WEF, 1993) and on energy conservation (University of Florida, 1986).

Energy consumption for a pump is dependent not only on the pump but on the hydraulics of the system in which it is installed. A pump may be rated at 2 700 m^3/d (500 gpm) at 18 m (60 ft) of head with 70% efficiency, but it will not discharge 2 700 m^3/d (500 gpm) at 18 m (60 ft) and 70% efficiency unless the hydraulics of the suction and discharge sides of the pump are matched exactly with the pump. Any deviation from those conditions will mean that the pump will perform less efficiently even though it may actually produce more discharge or more head.

To determine the actual discharge, head, and efficiency of a pump, it is necessary to understand two types of curves: pump characteristic curves and system head curves. These curves can be used to determine the operating point of the pump—the discharge, head, and efficiency.

PUMP CHARACTERISTIC CURVES. Pump characteristic curves are a reflection of the pump in new condition and are independent of the installation. Pump characteristic curves are plots of pump properties such as head, efficiency, power required, and net positive suction head required versus discharge. A typical set of curves is shown in Figure 5.1 and is relevant to a single pump with a single-sized impeller operating at a single speed. Because a given pump can usually accommodate a range of different impeller sizes, pump characteristic curves prepared by a manufacturer often show a family of curves for several impeller sizes, as shown in Figure 5.2.

The most important pump characteristic curves for energy conservation are the head characteristic and efficiency characteristic curves.

There are two important points on these curves: the best efficiency point (BEP) and the operating point. The BEP is described below. To understand the operating point, it is necessary to understand system head curves, so the description of operating point follows the discussion of system head curves.

BEST EFFICIENCY POINT. The BEP on a pump curve is the point corresponding to the maximum efficiency of the pump. For example, in Figure 5.1, the BEP corresponds to a flow of 5 500 m^3/d (1 000 gpm) with a head of 50 m (175 ft). The rated capacity (or duty point) of the pump is the discharge and head at (or near) the BEP. The BEP is a property of the pump and motor alone.

SYSTEM HEAD CURVE. The actual performance of a pump in any situation depends on the hydraulic conditions it encounters at any point in time. For wastewater pumps, this depends on a number of factors, including

- Water level on the suction side of the pump relative to the pump impeller centerline;
- Water level on the discharge side of the pump, if the pump discharges at or below the water surface (or the elevation at which the flow changes from full pipe to partly full on a free discharge);
- Head loss in the suction and discharge piping, which depends on the flows, diameters, and roughness of the pipes; and
- Influence of other pumps discharging into the same discharge line (whether they are in the same pumping station or other pumping stations).

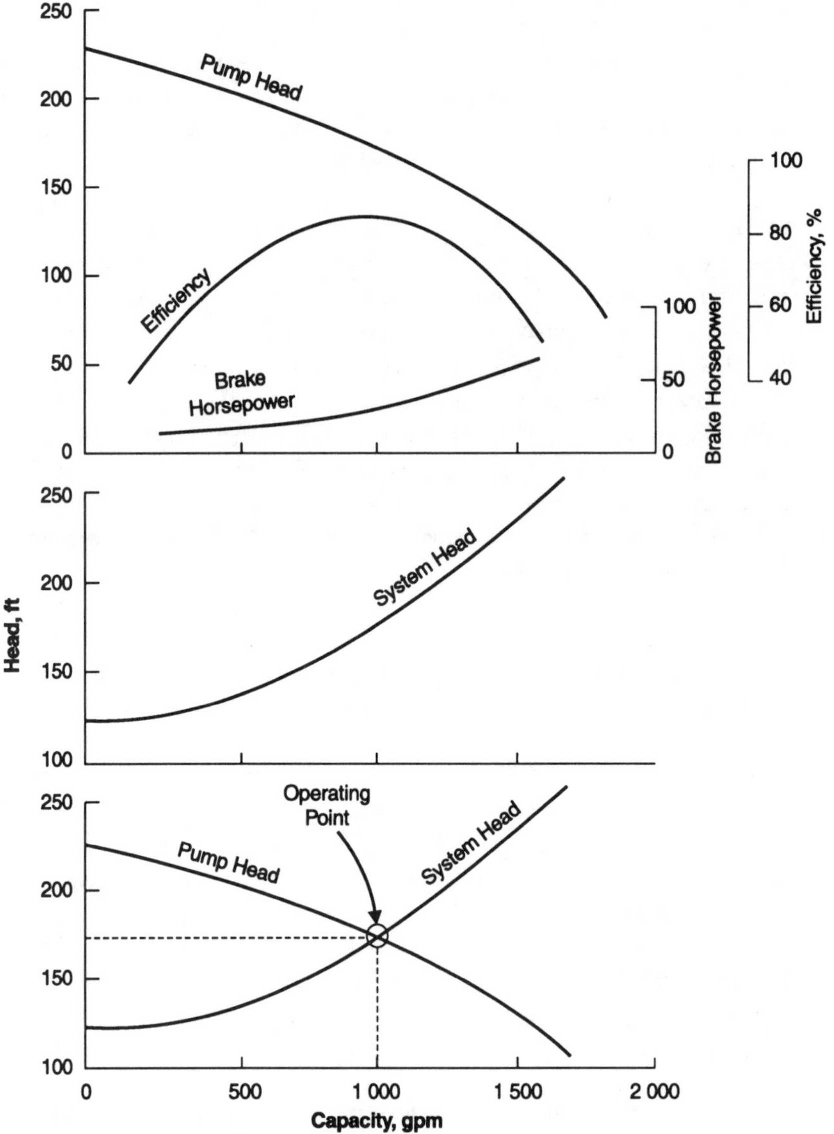

Figure 5.1 Typical pump characteristic curves (ft × 0.304 8 = m; gpm × [6.308 × 10^{-5}] = m^3/s; hp × 745.7 = W; hp × 745.7 = W) (reprinted with permission from University of Florida [1986] *Operations and Training Manual on Energy Efficiency in Water and Wastewater Treatment Plants.* Training Res. Educ. Environ. Occupations Cent., Gainesville, Fla.)

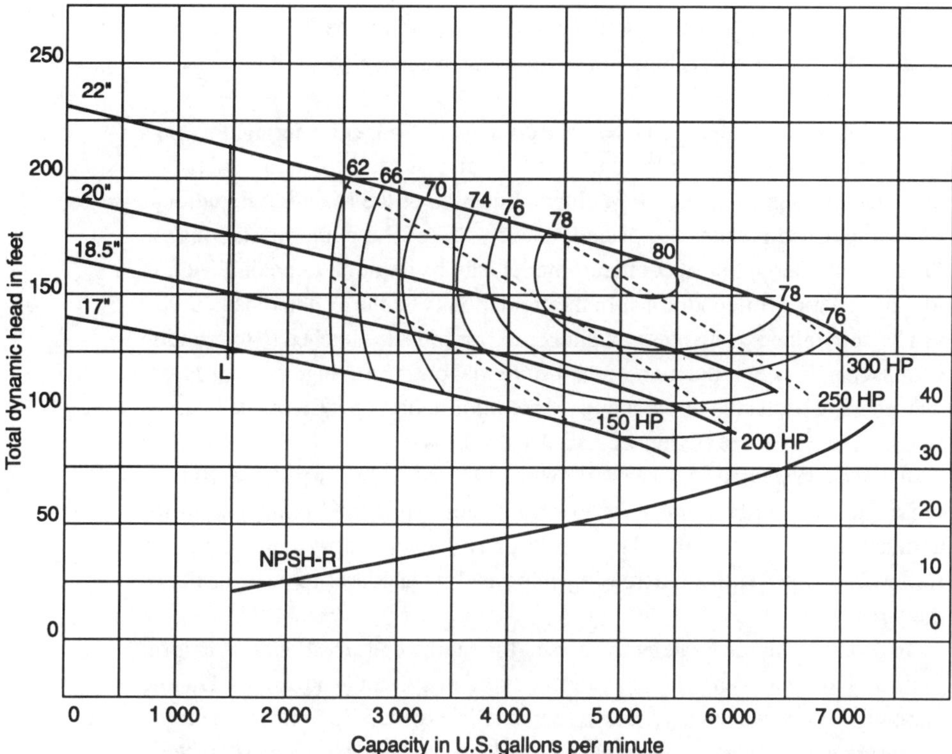

Figure 5.2 Pump characteristic curves for several impeller sizes (ft × 0.304 8 = m; gpm × [6.308 × 10^{-5}] = m^3/s

The relationship between the discharge of a pump and the head it must pump against is called the *system head curve.* In practice, there is not a single system head curve but, rather, a band of system head curves because the factors listed above can and do change over time. For example, tank levels rise and fall, other pumps turn on and off (or change speeds in the case of variable-speed pumps), valves can be throttled or closed, and pipe roughness and deposits can increase over time.

When tank levels remain constant and the status of other pumps in the same discharge line do not change, the system head curve can be represented by a single curve. The curve is always upwardly sloping because as pump discharge increases, head loss resulting from friction in the pipes also increases. For simple situations, the system head curve is the sum of the head required to lift the water from the suction tank to the discharge tank plus the head needed to overcome friction losses:

$$h = h_l + h_f \tag{5.1}$$

Where

h	=	system head, ft;
h_l	=	lift between suction and discharge point, ft; and
h_f	=	friction loss between suction and discharge point, ft.

The lift is easy to determine by measurement or from engineering drawings of the system. For simple pipelines, the friction loss can be determined using an equation such as the Hazen-Williams Equation or the Manning Equation. For complicated pipelines, especially those with several pumping stations on the same discharge line, a computer model may be required to predict friction loss for a variety of conditions. In the field, points on the system head curve can be determined using pressure gauges or transducers attached to the suction and discharge side of the pump or a differential pressure gauge or transducer connected to each side of the pump. More information on system head curves is available elsewhere (Ormsbee and Walski, 1989).

A typical system head curve is shown in Figure 5.1. In general, the lift term in Equation 5.1 will dominate for low flow rates, while the friction loss term dominates at high flow rates. This results in a relatively flat system head curve for low flow rates with an increasing slope as flow increases to the capacity of the piping.

In determining the strategy to be used for energy conservation, it is helpful to distinguish between different extreme types of system head curves. On one extreme is a flat system head curve (curve A in Figure 5.3), which is typical for systems with short, large pipelines and little friction loss, such as those found at influent pumping stations in WWTPs or in systems with pipelines that have been sized with considerable excess capacity. On the other extreme are steep system head curves (curve B in Figure 5.3), which are characteristic of systems with long force mains, high velocity, high friction losses, and little or no lift, typical of booster pump stations. Most system head curves lie somewhere between the two extremes. Energy conservation measures are considerably different for each extreme, as follows:

- For curve A, raising the wet well level can save some energy (for curve B, there would be little improvement); and
- For curve B, cleaning the discharge line by pigging or scraping can be helpful (pigging would result in no improvement for curve A).

For multiple variable-speed pumps discharging into the same force main, system head curves can vary widely. Figure 5.4 shows how system head curves can vary depending on whether other pumps in the same force main are running and whether they are running at high or low speed. System head becomes mostly a function of flow (velocity head). Consequently, pump efficiency will vary depending on the system head developed at the given time by

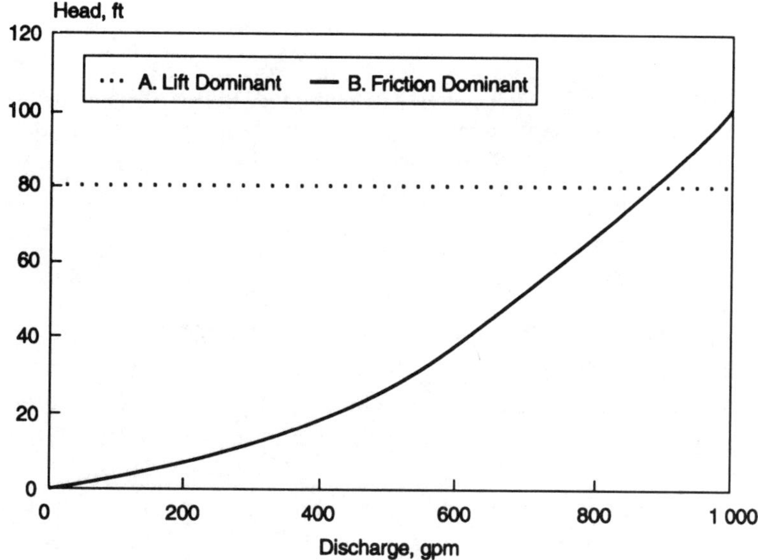

Figure 5.3 System head curves, example 1 (ft × 0.304 8 = m; gpm × [6.308 × 10^{-5}] = m^3/s)

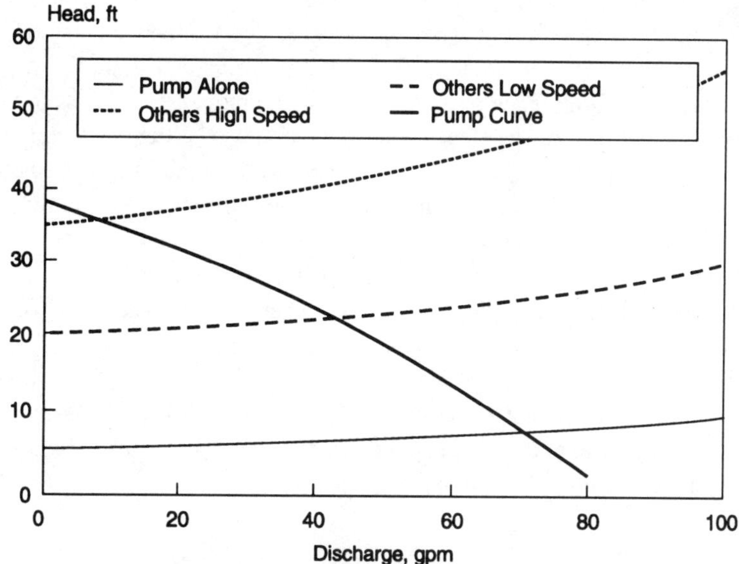

Figure 5.4 System head curves, example 2 (ft × 0.304 8 = m; gpm × [6.308 × 10^{-5}] = m^3/s)

the combination of pumps and resulting head, along with the pumps' characteristics, which have a definite head/efficiency relationship.

OPERATING POINT. The operating point is the point corresponding to the actual discharge and head produced by the pump. Graphically, it is the intersection of the pump head characteristic curve and the applicable system head curve at that point in time. The operating point depends on the pump, the flow, and the resulting hydraulics of the system in which it is installed and is shown graphically in the lower part of Figure 5.1.

RELATIONSHIP OF BEST EFFICIENCY POINT AND OPERATING POINT. Ideally, the operating point will be close to the BEP. Because the system head curve will continuously change, this will not always be possible. However, pumps should be selected and operated so that the operating point is reasonably close to the BEP.

When the BEP is not located near the operating point, it is necessary to investigate the cause of this difference. If the pump is not operating on its characteristic curve, air in the suction line, incorrect clearances, worn bearings, or even improper impeller size or motor speed are possible explanations. Typically, however, pumps perform on or near their characteristic curves.

If a pump is operating on its characteristic curve to the left (low-flow side) of its BEP, it is producing less flow than optimal. The pump is experiencing much more resistance to flow than it can handle. This can be the result of an undersized discharge line, some type of blockage (for example, partly closed valve) or encrustation in the suction or discharge piping, higher levels in the discharge tank (or lower levels in the suction tank) than anticipated, or a higher head in the discharge line than anticipated because of the operation of other pumps discharging to the same discharge line. The remedy is to reduce the resistance the pump must overcome.

If the pump is operating on its characteristic curve to the right (high-flow side) of the BEP, it is producing more flow than anticipated because it is experiencing less resistance to flow than anticipated. This can be because of an oversized discharge line, discharge or suction lines being cleaner than anticipated, low water level in the discharge tank (or high water level in the suction tank), or lower head in the discharge line than anticipated because of the operation of other pumps. Changing the impeller or pump operation may remedy this problem.

Regardless of the reason, the extent to which a pump operates at less than its BEP is a reflection of the extent to which the pump is mismatched with the hydraulics of the system it serves. This can be the result of improper selection of pumps, a change in operating conditions after the pump was installed, or simply expecting the pump to operate over too wide a range of conditions.

ENERGY PRINCIPLES

Energy is the ability to do work and produce power. Pumps convert some form of energy (mechanical, electrical) into water energy. The power required is defined in terms of work as weight of water per unit time (force) multiplied by total dynamic head (distance). A commonly used formula defining water power converted to units of horsepower (hp × 745.7 = W) is

$$\text{hp} = \frac{62.4 \times Q \times h}{550} \qquad (5.2)$$

Where

hp	=	water power, hp;
Q	=	flow rate, cfs;
h	=	pump head, ft;
62.4	=	specific weight of water, lb/cu ft; and
550	=	conversion from ft-lb/sec to hp.

Probably a more useful formula, which uses flow in million of gallons per day (mgd) and head in feet, is as follows:

$$\text{hp} = 0.175 \times Q \times h \qquad (5.3)$$

PUMP EFFICIENCY. The key to successful energy conservation for pumps is to produce water power using the smallest amount of electrical (or other) power as possible. The ratio of the water power produced by the pump to its power input is known as the overall pump efficiency, or the *wire-to-water* efficiency. The overall efficiency must consider the inefficiencies of the pump, the drive or transmission, and the power source such as motor or engine.

The mechanical efficiency of the pump is the ratio of water power from the pump to mechanical power input to the pump shaft. In formula terms for a direct-drive coupled motor,

$$e_p = \frac{0.113 \times Q \times h}{\text{hp}_\text{m}} = \frac{\text{water power}}{\text{motor power}} \qquad (5.4)$$

Where

e_p	=	pump efficiency;
Q	=	flow rate, cfs;
h	=	pump head, ft; and
hp_m	=	output of the motor, hp.

It is important to note that motors are rated by output rather than input. Therefore, a 75 000-W (100-hp) motor should deliver the full 75 000 W (100 hp) to the transmission or, in direct-drive applications, to the pump.

The motor and any drive mechanisms are not perfectly efficient. The efficiency of the motor can be determined as follows:

$$e_m = \frac{hp_m}{hp_e} = \frac{\text{motor power}}{\text{input power}} \tag{5.5}$$

Where

e_m = motor efficiency, and
hp_e = electric power (motor input power), hp.

Similarly, the efficiency of any associated transmissions must be considered and has a similar formula:

$$e_t = hp_p / hp_m \tag{5.6}$$

Where

e_t = transmission efficiency, and
hp_p = pump input power (or transmission output power), hp.

The overall efficiency of the pump and motor, referred to above as the *wire-to-water* efficiency, is the product of the pump efficiency, transmission efficiency, and motor efficiency:

$$e_{ww} = e_m \times e_t \times e_p \tag{5.7}$$

Where

e_{ww} = wire-to-water efficiency.

While efficiency is actually a number between zero and one, it is typically expressed as a percentage (that is, 80% rather than 0.80). In equations presented later in this chapter, the decimal form of the efficiency is used.

Typically, when pumps are tested by the manufacturer, the mechanical power input to the pump is controlled, and the resulting curves of efficiency versus flow are pump efficiency curves. In field applications, it is difficult to measure motor power input to the pump, so field testing of pumps almost always involves measurement of wire-to-water efficiency.

For example, a pump uses 1.1 kWh (4.0 MJ) of electricity in 10 minutes to pump 400 gpm (2 200 m^3/d) against a head of 55 ft (17 m). What is the wire-to-water efficiency of the pump and motor?

Calculate water horsepower:

$$Q(\text{cfs}) = 400 \text{ gpm} + 7.48 \text{ gal/cu ft} + 60 \text{ sec/min} = \frac{400 \text{ gpm}}{(448 \text{ gpm/cfs})}$$

$$\begin{aligned} \text{hp (water)} &= [(62.4 \text{ lb/cu ft})(0.891 \text{ cfs})(55 \text{ ft})]/(550 \text{ ft-lb/sec-hp}) \\ &= 5.56 \text{ hp} \end{aligned}$$

Calculate electric horsepower:

$$\begin{aligned} \text{hp (electric)} &= (1.1 \text{ kWh} \times 60 \text{ min/hr}) + (0.746 \text{ kW/hp}) + (10 \text{ min}) \\ &= 8.85 \text{ hp} \end{aligned}$$

$$e_{ww} = 5.56/8.85 = 0.63, \text{ or } 63\%$$

ENERGY CONSUMPTION. The energy consumption over a period of time is the water power multiplied by the length of time and divided by the wire-to-water efficiency:

$$E = \text{hp (water)} \times 0.746 \text{ kW/hp} \times t/e_{ww} \qquad (5.8)$$

Where

E = energy, kWh, and
t = time over which energy is consumed, hours.

The cost of energy can be determined by multiplying the energy used or forecast by the price of energy:

$$C = \frac{E \times P}{100} \qquad (5.9)$$

Where

C = cost of energy during time t, and
P = price of energy, cents/kWh.

Substituting Equation 5.8 for E,

$$C = \frac{[\text{hp (water)} \times 0.746 \text{ kW/hp} \times t/e_{ww}] \times P}{100}$$

Substituting Equation 5.2 for hp (water),

$$C = \frac{\frac{62.4 \text{ lb/cu ft} \times Q \times h}{550 \text{ ft-lb/sec/hp}} \times (0.746 \text{ kW/hp} \times t/e_{ww}) \times P}{100}$$

with flow given in gpm (448 gpm/cfs) gives

$$C = (1.89 \times 10^{-6}) \times Q(\text{mgd}) \times h(\text{ft}) \times t(\text{hr}) \times P(\text{cents/kWh})/e_{ww}$$

or with flow given in mgd (0.646 gpm/mgd) gives

$$C = 0.001\,31 \times Q(\text{mgd}) \times h \times t \times P/e_{ww}$$

The annual cost (AC) of energy can be determined by inserting 8 760 hours/year for t to yield

$$\text{AC} = 0.016\,5 \times Q(\text{gpm}) \times h \times P/e_{ww} \tag{5.10}$$

or for large pumps, whereby flow is measured in million of gallons per day

$$\text{AC} = 11.4 \times Q(\text{mgd}) \times h \times P/e_{ww} \tag{5.11}$$

Where

AC = annual cost of energy, \$/yr.

Find the expected annual energy cost to operate a raw water intake pump at a WWTP that, on average, pumps 24 mgd (1.1 m^3/s), with 40 ft (12 m) of head, an overall efficiency of 60%, and an energy cost of 8 cents/kWh (kWh × 3.600 = MJ):

$$\begin{aligned} \text{AC} &= 11.4 \text{ (24 mgd)(40 ft)(8 cents/kWh)}/0.60 \\ &= \$146\,000\text{/yr.} \end{aligned}$$

MULTIPLE OPERATING POINTS. In most cases, it is not sufficiently accurate to use the average flow, head, and efficiency of a pump (or station) to calculate energy cost. Instead, the values corresponding to each operating point should be calculated and summed. If there is an operating point possible for a pump or group of pumps at a pumping station during time period of length T, then energy cost can be calculated as

$$C = 1.89 \times 10^{-6} \; P \sum_{i=1}^{n} Q_i(\text{gpm}) \; h_i t_i / e_{wwi} \tag{5.12}$$

Where

$$\sum_{i=1}^{n} t_i = T \tag{5.13}$$

i = subscript of operating point, and
n = number of operating points.

For example, a pumping station can operate with zero, one, two, or three identical pumps operating at any time. Over the course of a year, the station operates as shown in the first five columns of Table 5.1. Find the energy cost if the price of power is 9.2 cents/kWh.

Table 5.1 Energy cost example for pumping stations

Pumps running	Q_i	h_i	t_i	e_i	$1.89 \times 10^{-6} Qht/e$
0	0	0	240	—	—
1	900	45	6 260	62	773
2	1 700	48	2 120	61	535
3	2 500	55	140	57	64
Sum					1 372 × 9.2 = \$12 622

As a quick approximation, round off and assume that, on average, the pumping station pumps 1 100 gpm, with 46 ft of head and an efficiency of 61%, and use Equation 5.10 to give

$$\text{AC} = 0.016\,5\,(1\,100)(46)(9.2)/0.61 = \$12\,600$$

Now that the equations for pumping energy have been presented, it is possible to examine every term (Q, h, P, e) to find ways to reduce energy consumption. The following sections present methods to conserve energy.

DISCHARGE

Reducing pump discharge is the most direct way to reduce energy consumption. Energy savings as a result of reductions in flow are typically proportional to the amount of flow reduced, and sometimes even greater. There are often opportunities to realize savings by minimizing the flow, for example, by reducing recycle streams and infiltration and inflow into the collection system.

For constant-speed pumps, reductions in flow typically are realized at a pumping station as a reduction in the amount of time the pumps actually run, rather than appearing as a reduction in instantaneous flow rate while the pumps are running. This means that for a single pump on a pipeline, energy savings should be proportional to flow reduction. When multiple pumps are involved, the pumps will operate further to the right on the system head curve for longer periods of time, so the savings will be more than proportional.

For variable-speed pumps, reduction in flow to the pumping station appears as a decrease in the flow and head produced by the station. It will rarely reduce the duration for which the pumps run. Depending on the pump efficiency curve, reductions in flow may result in decreased pump efficiency. Nevertheless, the major savings will be from the reduced flow.

HEAD

Reducing the head that a pump must work against results in a proportional reduction in energy consumption, provided the efficiency of the pump and motor do not change significantly. Because head is the result of lifting the water or overcoming friction, reduction in head can be achieved by reducing the lift or minimizing friction losses.

MEASURING HEAD. The best way to measure the head that a pump is producing is to measure the head with a differential pressure gauge with one tap immediately upstream and the second tap immediately downstream of the pump. For greatest precision, the gauge should have a range slightly greater than the anticipated shutoff head of the pump. The change in velocity head should be added or subtracted from the pump head reading. In most cases, however, the change in velocity head across the pump is negligible.

If a differential gauge is not available, then a pressure gauge for the discharge and a vacuum (suction) gauge should be used. If the discharge gauge is at a higher/lower elevation than the suction gauge, the difference in elevation of the gauges should be added to/subtracted from the pump head.

On occasion, a convenient location to tap the pipe near the pump is not available, and it becomes necessary to use a pressure reading at a more distant location to determine the suction or discharge pressure. In such a situation, it is necessary to estimate the head loss in pipe and fittings between the measured head and the pump and the change in velocity head, and subtract these heads from the measured suction head (or add these values to the measured discharge head).

SYSTEM HEAD. In the case of a single pumping station discharging to a force main, there is little that can be done to reduce the discharge head. However, controlling the flow with variable-speed pumps and minimizing multiple-pump or peak flow operation will minimize the head losses resulting from friction, which is especially important in systems with long mains.

When multiple pumping stations discharge into a force main, it is important to ensure that the pumps are evenly matched. Otherwise, one or two pumps with high heads can increase pressures in the force main so that the higher head pumps back up and possibly even shut off the lower head pumps (shutoff head is the point at which the pump characteristic curve intersects the vertical axis—that is, discharge equals zero). Because all of the wet wells will be at slightly different elevations, it is important to make these comparisons based on discharge hydraulic grade line relative to the same datum point (for example, sea level) and not simply make comparisons based on discharge head or pressure. This is especially true for constant-speed pumps, but problems can even be experienced in variable-speed stations.

For example, given that there are four pumping stations discharging to a force main, determine the discharge head for each and determine whether they will work well together. Data on the pumping stations are given in the first four columns of Table 5.2. The last two columns are the sum of the average wet well level and the appropriate pump head. Within each pumping station, all pumps are identical.

Table 5.2 Data for four pumping stations discharging to a force main

Pump station	Average wet well elevation	Head at BEP[a]	Shutoff head	Head at BEP	Head at shutoff
	ft[b]	ft	ft	ft	ft
1	520	45	60	565	580
2	505	55	75	560	580
3	500	80	110	580	610
4	490	65	85	555	575

[a] Best efficiency point.
[b] ft × 0.3048 = m.

In this example, pumping station 3 is not well matched with the other stations. It would tend to use up the capacity of the force main, and if the force main does not have a large capacity, it would cause the other pumping stations to run inefficiently. If the force main has adequate capacity and pumping station 3 has variable-speed drives, the combination in the example may be acceptable. However, it is best to avoid such a situation.

FRICTION. Friction losses are inevitable whenever fluid flows, but they can become excessive and result in increased energy use. Situations in which normal friction losses can increase include

- Lack of capacity of force main,
- Excessive roughness in force main, and
- Blockage of force main with debris or partly closed valve.

DETERMINING HEAD LOSS. The friction energy loss is the sum of the loss in straight runs of pipe, which can be determined from the Hazen-Williams, Manning, or Darcy-Weisbach equations, plus minor losses resulting from valves, bends, and fittings, which can be determined from hydraulic handbooks or manufacturers' literature. For long pipelines, minor losses are usually negligible compared with line losses, but for short lift lines at WWTPs, minor losses can be of greater significance.

In English units, the Hazen-Williams equation can be written as follows (see Appendix for conversions to metric):

$$h = (10.4\ L/D^{4.87})(Q/C)^{1.85} \tag{5.14}$$

Where

h	=	head loss, ft;
L	=	length of pipe, ft;
D	=	diameter, in.;
Q	=	flow, gpm; and
C	=	Hazen-Williams C-factor.

The C-factor is an indicator of the carrying capacity (smoothness) of the pipe and can range from 140 for new, smooth pipe to 30 for old encrusted pipe. For design, values of 80 to 100 are typically used.

Minor losses for valves and fittings are usually converted to length of straight pipe, which can be added to actual lengths of straight pipe to estimate overall head loss. For example, for a 0.15-m (6-in.) pipe with a fitting that would result in a loss of 6 diameters, the fitting would cause the same head loss as 0.9 m (3 ft) of pipe. One would need to add 0.9 m (3 ft) to the length used in the straight pipe head loss formula to account for the losses caused by the fitting. Equivalent pipe length conversion values for common valves and fittings can be found in many books, handbooks, and other references (ASCE, 1992; Sanks, 1989; Walski, 1984).

A quick way to determine whether capacity is adequate is to check the velocity in the pipe during normal operation. The velocity should be greater than 0.6 m/s (2 ft/sec) to ensure scour and less than 1.5 m/s (5 ft/sec) under normal flow conditions. During peak flow conditions, the velocity should be less than 3 m/s (10 ft/sec). If these velocities are being exceeded, the force main may be too small. The pipe may have been undersized initially, or more customers may have come on line recently, exceeding the capacity of the pipe.

Another problem is that the individual pumps may be too large for the force main, as illustrated in the following example.

A 2 000-ft (600-m) long 6-in. (0.15-m) force main with a C-factor of 100 carries an average flow of 100 gpm (550 m^3/d) and an expected peak flow of 400 gpm (2 200 m^3/d) (minor losses are negligible). First, determine the system head curve, given that the average lift is 20 ft (6 m). Second, determine the average operating point for the two different constant-speed pumps (A and B) given in Figure 5.5. Third, for each pump, determine the percentage of time and number of hours the pump will run during the year. Finally, determine the energy cost for each pump if the wire-to-water efficiency for each is 60% and the price of energy is 8 cents/kWh.

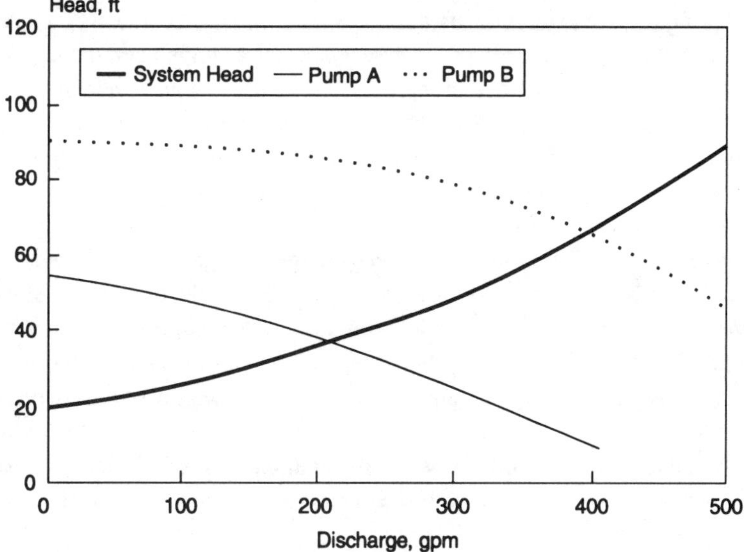

Figure 5.5 Pump comparison (ft × 0.304 8 = m; gpm × [6.308 × 10^{-5}] = m^3/s)

The system head curve can be determined by inserting $D = 6$, $L = 2\,000$, and $C = 100$ into the Hazen-Williams equation and adding the lift of 20 ft (6 m) to yield the following:

$$\begin{aligned} h &= [10.4(2\,000)/6^{4.87}](Q/100)^{1.85} + 20 \\ &= 0.000\,675\,Q^{1.85} + 20 \end{aligned}$$

The system head curve generated from the above equation is plotted on Figure 5.5.

From Figure 5.5, the operating point for pump A is 200 gpm (1 100 m^3/d) at 38 ft (12 m) of head; for pump B, it is 400 gpm (2 200 m^3/d) at 64 ft (20 m) of head. Because the average flow is 100 gpm (550 m^3/d), pump A will run

50% of the time, or 4 380 hours per year, while pump B will run 25% of the time, or 2 190 hours per year.

The annual energy consumption for each pump can be determined as

$$\begin{aligned}\text{Cost (A)} &= [(1.89 \times 10^{-6})(200\text{ gpm})(38\text{ ft})(4\,380\text{ hr/yr})(\$0.08/\text{kWh})]/0.6 \\ &= \$839/\text{yr}\end{aligned}$$

$$\begin{aligned}\text{Cost (B)} &= [(1.89 \times 10^{-6})(400\text{ gpm})(64\text{ ft})(2\,190)(\$0.08/\text{kWh})]/0.6 \\ &= \$1\,413/\text{yr}\end{aligned}$$

The above example illustrates that even though each pump was selected with the same efficiency, energy price, and total annual discharge, the pump that produced the lower friction head loss resulted in greater energy savings. This is because the velocity in the pipe was twice as high when the larger pump (B) ran. The smaller pump (A) still produced adequate velocity (2.3 ft/sec [0.7 m/s]) when it ran. If the utility was concerned about infrequent peak flows and wanted greater peak capacity, it could achieve this by adding another parallel pump like A rather than using pump B.

LOW FLOWS. One situation in which energy savings can be realized occurs when a pump has been sized for future growth but, in the early years of its operating life, the flows are considerably less, so the pumping station is not running most of the time. In such a case, the user can save money by shaving the impeller or replacing the impeller with a downsized impeller, saving the original impeller for when flows increase. The smaller impeller can pay for itself in a few years provided it is not so small that loss of efficiency is excessive or that the velocities in the force main fail to reach scour velocity.

PIPE RESTRICTIONS. In addition to a force main being undersized to carry the required flow, excessive head loss can also be a result of some type of restriction in the flow. It is desirable to record discharge pressure from pumping stations. If that pressure should increase abruptly, then it is possible that someone has left a valve partly closed or some large object has entered the discharge pipe. An abrupt drop in suction head can result from a partly closed valve or a foreign object entering the line on the suction side of the pump.

If the increase in discharge head or decrease in suction head occurs gradually over time, then sedimentation in the pipe, scaling, corrosion, or air blockage are more likely reasons for a rise in discharge head and/or a decrease in discharge flow rate. Checking air-release valves or installing valves at high points along the force main can correct problems with air blockage. Sedimentation or encrustation can best be corrected by cleaning the line by pigging or scraping.

Cleaning the force main increases the carrying capacity, which moves the system head curve to the right, increases pump discharge, decreases pump head, and generally saves energy. An exception to this is, for example, if a pump has been sized to pump into a discharge line with a low carrying capacity, and the carrying capacity actually turns out to be much higher, as illustrated in the following example.

A pump is discharging into a 2-mile (3.2-km) force main with a Hazen-Williams C-factor of 80 and a diameter of 8 in. (0.20 m). After cleaning of the pipe by pigging, the C-factor was increased to 110. The lift head is negligible compared to the head required to overcome friction losses. The pump head and wire-to-water efficiency curves are given in Figure 5.6. Plot the system head curve, determine the operating points for both the original and cleaned line, determine the number of hours each pump must run during the year to pump an average 150 gpm (of 820 m^3/d), and determine the annual energy cost for each operating point based on an energy cost of 7 cents/kWh.

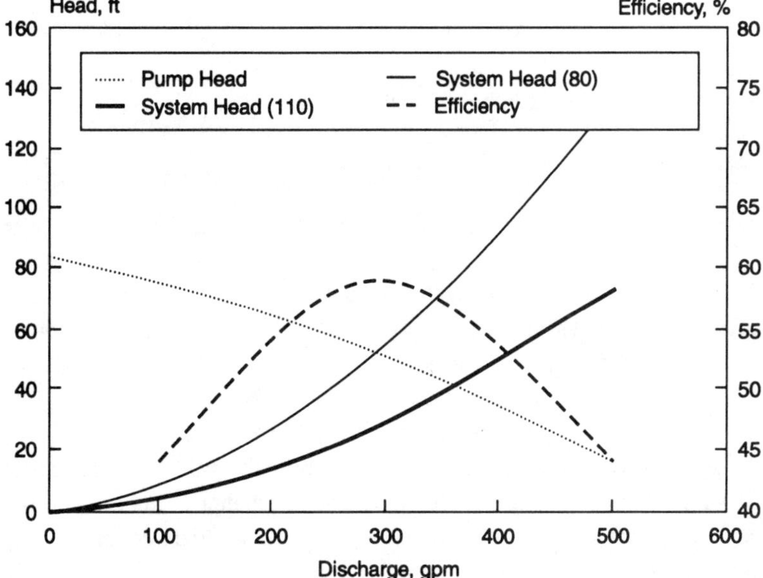

Figure 5.6 Changing C-factor (ft × 0.304 8 = m; gpm × [6.308 × 10^{-5}] = m^3/s)

The Hazen-Williams equation is used to develop the system curve as follows:

$$h = [(10.4)(10\,560)/8^{4.87}](Q/C)^{1.85}$$

For $C = 80$, the equation simplifies to

$$h = 0.001\,32 \times Q^{1.85}$$

For $C = 100$, the equation is

$$h = 0.000\,734 \times Q^{1.85}$$

The system head curves are plotted in Figure 5.6. They show that for $C = 80$, the operating point corresponds to a flow of 300 gpm (1 600 m^3/d), a head of 50 ft (15 m), and a wire-to-water efficiency of 60%. For $C = 110$, the operating point corresponds to a flow of 360 gpm (2 000 m^3/d), a head of 40 ft (12 m), and an efficiency of 55%.

Because the average flow over the course of a year is 150 gpm (82 m^3/d), the number of hours each pump must run can be determined as follows:

$$t = (150/300)(8\,760) = 4\,380 \text{ hours for } C = 80$$
$$t = (150/360)(8\,760) = 3\,650 \text{ hours for } C = 110$$

The annual energy cost can be calculated as

$$AC = [(1.89 \times 10^{-6})(300)(50)(4\,380)(7)]/0.6 = \$1\,448/yr$$
$$AC = [(1.89 \times 10^{-6})(360)(40)(3\,650)(7)]/0.55 = \$1\,264/yr$$

The example above shows that the annual energy savings was $184/year by pipe cleaning. The reason the savings were not greater is that the efficiency of the pump dropped because it was sized to pump against the larger head. If, on the other hand, the pump were at a maximum efficiency of 60% at $C = 110$ (AC = $1 159/yr) and up from an efficiency of 55% when $C = 80$ (AC = $1 580/yr), the annual energy savings by cleaning the pipe would have been $421/year, which is substantially greater. The key is that it is necessary to check the operating points when attempting to estimate the benefits of a pipe-cleaning project.

FLOW METER CALIBRATION VERIFICATION

Field measurement of flows is difficult to perform with great accuracy. Numerous techniques are available for measuring flow, with new devices being introduced to the market each year (see U.S. EPA, 1981, or Walski, 1984, for more information). Wastewater can coat sensors or plug gauges, and pipe walls can become coated with material. All of these factors can reduce the accuracy of flow-measuring devices, which means that constant maintenance and routine calibration are necessary.

Because flow is defined as the volume passing a point at a given time, the most sound way to verify the calibration of a flow meter is to measure the change in volume in some kind of vessel over a period of time. This can be done in a pumping station wet well and is referred to as a *drawdown* test. To perform such a test, it is necessary to measure the plan area of the wet well (subtracting out large objects) and accurately measure the change in water level (with a staff gauge or pressure transducer). The primary source of error is inflow to the wet well during the test. If such flow is small, it can be ignored. If it is larger, the inlet can be plugged for a short period of time. If it is larger still, the inflow can be measured for a period of time with the pumps off, and an estimate of the inflow rate can be added to the pump discharge.

For example, a wet well measuring 10 ft × 20 ft (3 m × 6 m) with approximately 8 sq ft (0.7 m^2) taken up by equipment and columns. Inflow was estimated at 41 gpm (223 m^3/d) with the pumps off. With one 400 gpm (2 200 m^3/d) pump running, the water level dropped by 2.25 ft (0.67 m) during a 10-minute test. Find the pump discharge:

$$\begin{aligned}\text{Discharge} &= \text{Volume change/Time} + \text{Inflow}\\ &= 2.25\text{ ft} \times [(10\text{ ft} \times 20\text{ ft}) - 8\text{ sq ft}] \times (7.48\text{ gal/cu ft/10 min}) + 41\\ &= 323\text{ gpm} + 41\text{ gpm} = 364\text{ gpm}\end{aligned}$$

REFERENCES

American Society of Civil Engineers (1992) *Pressure Pipeline Design for Water and Wastewater.* Committee Pipeline Plann., New York, N.Y.

Hicks, T.G., and Edwards, T.W. (1971) *Pump Application Engineering.* McGraw-Hill, Inc., New York, N.Y.

Hydraulic Institute (1983), *Hydraulic Institute Standards for Centrifugal, Rotary and Reciprocating Pumps.* 14th Ed., Cleveland, Ohio.

Karassik, I.J., *et al.* (1976) *Pump Handbook.* McGraw-Hill, Inc., New York, N.Y.

Ormsbee, L.E., and Walski, T.M. (1989) Developing System Head Curves for Water Distribution Pumping. *J. Am. Water Works Assoc.,* **81,** 7, 63.

Pincince, A.B. (1970) Wet-Well Volume for Fixed Speed Pumps. *J. Water Pollut. Control Fed.,* **42,** 1, 126.

Sanks, R.L. (Ed.) (1989) *Pump Station Design.* Butterworths, Boston, Mass.

University of Florida (1986) *Operations and Training Manual on Energy Efficiency in Water and Wastewater Treatment Plants.* Training Res. Educ. Environ. Occupations Cent., Gainesville, Fla.

U.S. Environmental Protection Agency (1981) *NPDES Compliance Flow Measurement.* MCD-77, Washington, D.C.

Walski, T.M. (1984) *Analysis of Water Distribution Systems.* Krieger Publishing, Malabar, Fla.

Water Environment Federation (1993) *Design of Wastewater and Stormwater Pumping Stations.* Manual of Practice No. FD-4, Alexandria, Va.

Yin, M.T., *et al.* (1996) Optimum Simulation and Control of Fixed Speed Pumping Stations. *J. Environ. Eng.,* **122,** 3, 205.

Chapter 6
Variable Controls

A variety of methods is used to control the output of pumps, blowers, and other powered equipment. These methods range from direct control of the driver (motor or engine), to control of the speed or force between the driver and the driven element, to control of the driven element itself. In any case, the control method itself introduces another device, with its own level of inefficiency, that must be taken into consideration. Variable controls are used for several reasons, but often they do not lead to greater system efficiency.

Most often, variable controls are used to provide a continuous, noninterrupted output, which varies with either the demands of the input or the demands of the output as imposed by the system or the operator. In the case of a main wastewater pump at a wastewater treatment plant (WWTP), it is generally better to have a uniform flow rather than to have a single pump abruptly turning on and off and causing surges through the system. However, in larger WWTPs using multiple pumps, it is more efficient to use constant-speed pumps for the base load, with one variable-controlled pump adjusting for the changing flow conditions.

One of the exceptions to increased efficiency using variable control is the case of a pumping station with a high total dynamic head (see Chapter 5) caused more by friction losses than by elevation head (for example, flow through long, narrow pipelines). Because the friction head loss is often a function of the square of the velocity (and, therefore, square of the flow), greater inefficiency may result from the friction losses caused by full-pump flow than from the variable-control device.

TYPES OF VARIABLE CONTROLS

INDIRECT CONTROL. The output of a pump, to a minor degree, self-regulates to the demands being placed on it. For example, as wet well elevation varies, a centrifugal pump's output will vary. Its discharge will increase as intake elevation increases, and its discharge will decrease as intake elevation decreases. While this may not provide the degree of operator control desired, it must be taken into account in pumping station design.

CONTROL OF DRIVER Motors can be controlled directly by various electrical means. Direct current (dc) motors can be controlled with rheostats (variable resistance), which decrease the current to the motor, resulting in decreased power to the motor. Alternating current (ac) motors can be equipped with electrical packages that provide variable voltages or variable frequencies. In recent years, preference has been leaning toward variable-frequency drives because of their high efficiency, good reliability, and reasonable cost. Solid-state inverters are currently used for varying the voltage or frequency to motors to provide variable output. They can be added to existing induction motor installations without any modifications to the pump/motor system. Most often, inverters are supplied with their own enclosures, which are mounted near the motor starter.

Engines are controlled simply by varying the supply of fuel: the more fuel supplied, the greater the power produced. The automobile accelerator is an example of how the fuel supply controls engine speed.

CONTROL BETWEEN DRIVER AND DRIVEN ELEMENT. A variable-control device located between the driver and the driven element is called a *drive*. The drive can be used to change the speed of the shaft of the driven element or provide a mechanical advantage by changing the force (torque) of the driven element. Drives can be mechanical, such as transmissions, gear boxes, pulleys, or sheaves; electromechanical, such as magnetic slip clutches; or electronic, such as variable voltage or variable frequency.

CONTROL OF DRIVEN ELEMENT. The output of driven elements such as centrifugal pumps can be varied or controlled simply by throttling a valve on the discharge side of the pump. Likewise, a centrifugal blower's output can be varied by throttling the intake of the blower. While effective, throttling is inefficient because the pump or blower fights the resistance even though it still does less work.

Among speed-control drives, the ones with highest efficiencies are the variable-frequency and dc drives, while the belt-driven, constant-speed motor with adjustable sheaves is the cheapest.

BLOWERS AND COMPRESSORS: CONTROLS AND DRIVES

POSITIVE DISPLACEMENT BLOWER. Positive displacement blowers deliver a given volume of air at a relatively fixed discharge pressure and given speed. Therefore, a positive displacement blower should never be throttled. Its power requirement is directly related to the volume and density of air moved for the discharge pressure encountered. Accordingly, the common method for controlling output and energy consumption is through a change in the blower speed.

Centrifugal Blower. A centrifugal blower is typically designed to meet one particular set of conditions at its best efficiency point. This particular performance can seldomly accommodate the range of flows and pressures required for a WWTP. To meet lower capacity conditions, a method of capacity turndown or regulation must be employed. Centrifugal blowers deliver a volume of air as the rotating impeller imparts a centrifugal force on the gas. The power required by a centrifugal blower is proportional to the weight of air moved and the discharge pressure. The air flow rate can be controlled via the blower speed, inlet throttle valve, or inlet guide vanes.

Intake throttling has commonly been used on blowers as an expedient, low-cost method for capacity turndown. While reducing energy usage, intake throttling is not necessarily efficient. As the intake of a blower is throttled, a resulting increase in suction reduces the density of the feed air, which in turn reduces the air mass flow rate. Nevertheless, many centrifugal blowers are equipped with either an adjustable inlet guide vane or a simple butterfly valve to throttle the inlet gas and effect variable blower output. The adjustable inlet guide vane is a more efficient device for capacity regulation than a simple inlet valve ahead of the blower. The adjustable inlet guide vane prerotates the incoming gas, which results in a reduction of both capacity and discharge pressure. All blowers have limited turndown capacity. If the blower is turned down too low, surging results from insufficient flow through the blower, leading to noises, vibrations, and heating of the air. If surging is allowed to persist, damage will result to the blower and possibly to related equipment.

Adjustment of the inlet guide vanes can reduce the surge limit to 30% of the rated capacity, whereas throttling with inlet valves can reduce the surge limit to no lower than 45% of the rated capacity.

For a centrifugal blower equipped with a radial-type impeller, the adjustable discharge diffuser can provide turndown regulation without reducing the discharge pressure. Stable operation without surging can be attained down to 45% of the rated capacity or less. When used in conjunction with inlet throttling by guide vane adjustment, stable regulation to flows of approximately 30% of the rated capacity is possible.

Chapter 7
Energy Utilization in Wastewater Treatment Processes

Energy use may vary significantly for a given size wastewater treatment plant (WWTP), depending on its location, strength of wastewater, level of treatment, in-plant recovery, type of treatment process selected, and mode of operation. Therefore, it is not practical to provide a general energy usage level for WWTPs. Generally, conventional aquatic lagoon systems and land treatment systems are among the least energy intensive. Advanced treatment processes often require greater amounts of primary energy and also require greater amounts of chemicals, with associated secondary energy use. Factors such as wastewater strength, hydraulic conditions, in-plant energy-recovery design, and operation mode greatly affect electrical power and fossil fuel energy uses. Typical profiles for annual energy use are shown in Figure 7.1 for different types of secondary treatment WWTPs with various treatment processes and at 10 mgd (0.4 m^3/s) average flow.

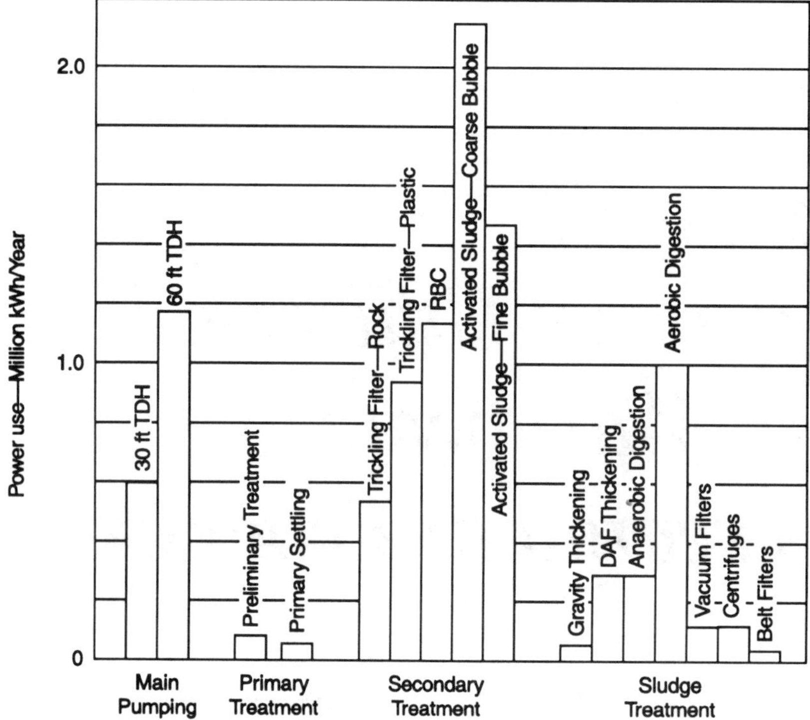

Figure 7.1 Typical energy use profile for 10-mgd (0.4-m^3/s) secondary treatment processes

The energy uses of WWTPs are generally considered to be those directly associated with on-site electrical power and fuel requirements. These are typically referred to as *primary* energy uses. Secondary energy uses are associated with materials manufacture and transport and manufacture of consumable chemicals that are used in the treatment process. A discussion of energy use for specific unit processes follows.

PRELIMINARY AND PRIMARY TREATMENT

SCREENING AND COMMINUTORS. Screens are usually the first treatment devices in WWTPs. Screening units are cleaned either manually or by mechanical chain-driven raking devices. Mechanically cleaned bar screens are used for all but small WWTPs. Screening represents a minor portion of total WWTP energy use, with units handling 0.7 m^3/s (15 mgd) or less typically being driven by 550-W (0.75-hp) motors. In general, it is not necessary to operate the cleaning rakes continuously, and most applications include adjustable timing devices or differential water level devices to activate the screen-cleaning mechanism. In small WWTPs, comminutors are used for shredding and grinding screenable debris to reduce particle size. Energy required for comminutor devices is 0.2 $W/m^3{\cdot}d$ (1.0 hp/mgd) wastewater flow. Comminutors and barminutors return the ground debris back to the wastewater, which adds to WWTP energy demands in subsequent processes.

An energy conservation measure is as follows:

- When practical, remove screenings from the WWTP and dispose of them in a landfill rather than grinding them and returning them to the treatment process.

INFLUENT WASTEWATER PUMPING. Depending on the WWTP site elevation and influent sewer elevation, the influent wastewater pumping energy requirement alone can represent 15 to 70% of the total WWTP use of electrical energy. Moreover, if the energy required to operate all of the pumps in the collection system is considered, total pumping energy requirements may represent as much as 90% of the total energy used. Actual power consumption by a pumping system may be expressed as follows:

$$\mathrm{hp_w} = \frac{0.175\,5 \times Q \times h}{e_T} \quad \text{or} \quad \mathrm{kW_w} = \frac{0.130\,9 \times Q \times h}{e_T} \qquad (7.1)$$

Where

$\mathrm{hp_w}$	=	motor input power for pump (wire hp), hp;
$\mathrm{kW_w}$	=	motor input power for pump (wire kW), kW;
Q	=	pump flow, mgd;
h	=	pump total dynamic head (TDH), ft; and
e_T	=	$e_p \times e_d \times e_m$; e_T is overall operating efficiency of the pumping system; e_p, e_d, and e_m are operating efficiencies of the pump, drive, and motor, respectively, in decimal equivalents.

For a typical direct-drive influent pump with a TDH of 9 m (30 ft), a pump efficiency of 74%, and a motor efficiency of 88%, the pump wire-to-water power requirement would be 1.6 $W/m^3{\cdot}d$ (8.1 hp/mgd) or 6.0 kW/mgd (mgd × [4.383×10^{-2}] = m^3/s) (see Chapter 2). Estimating power requirements by multiplying the influent flow received by the pump system energy ratio is much more accurate than using the motor horsepower and estimated run times (see Chapter 5 for a more complete description of energy requirements for pumping).

As an example, assume that a station with two 100-hp (75 000-W), 12-mgd (0.5-m^3/s) pumps and a design TDH of 30 ft (9 m) processes 18.0 mgd

(0.8 m^3/s) over a 24-hour period. What is the estimated average power draw and energy requirement?

$$\text{Average power} = 18.0\ \text{mgd} \times 6.0\ \text{kW/mgd} = 108\ \text{kW} \quad (7.2)$$

$$\text{Energy required} = 108\ \text{kW} \times 24\ \text{hr} = 2\,592\ \text{kWh/day} \quad (7.3)$$

GRIT REMOVAL. Grit removal is not a process that consumes a great deal of energy; nevertheless, there is some energy use and thus some room for conservation. In velocity-type grit basins, little energy is used, and once the basin is constructed, not much additional energy can be saved. In aerated grit chambers, the blowers do use a considerable amount of energy, so that reducing the air produced by the blowers could result in reductions in energy use.

The key to optimal operation of aerated grit chambers is matching the air produced with the air required for optimal operation. Too little air will result in putrescible material settling at the bottom of the tank, producing odors and making storage and disposal of grit more difficult. Too much air will result in too little grit being removed. This puts a greater burden on downstream unit processes and sludge disposal operations. Grit that might have been hauled to a landfill must pass through all of the sludge-processing operations.

Matching blower output with necessary air requirements is a trial-and-error process. There is no theoretical way to exactly determine the blower output required. The optimal output can only be determined by repeated tests. The best criterion for judging whether the air flow is correct is to examine the grit removed. If it only contains large, coarse particles, the air flow can be cut back until putrescible light organic matter begins to show up in the grit. Conducting a grain-size distribution test is a rigorous way of confirming visual inspection of the grit.

In installations with variable-speed drivers for the blowers, matching the blower output with air requirements is easy. However, variable-speed grit blowers are seldom used. In some cases, the blowers are driven by belts. In these instances, changing the sheaves can produce optimal air flow. This is considerably more time consuming than adjusting a variable-speed motor, but it is still worthwhile.

If changing the air flow requires changing the driver, an economic analysis should be done to determine whether this change is justifiable. The best solution in this case may be to throttle the air flow with valves. This may help match the air required with the air provided. However, because of the additional head loss in the air manifold, such throttling may or may not reduce energy use. This needs to be checked with a power meter at different air flow rates.

PRIMARY TREATMENT. The purpose of primary treatment is to remove settleable solids and floating material from the wastewater. Primary settling tank energy uses include drives for collector mechanisms, primary sludge pumping, and floatable skimmings handling. The primary sludge pumping operation is generally the largest energy-consuming component of the primary treatment process. A hidden cost of primary treatment is the ensuing cost of secondary treatment, which is affected by removals in primary treatment. Poor primary removals may require greater use of aeration energy in the secondary process and possibly more energy in solids handling and disposal as a result of the production of a greater secondary-sludge–to–primary-sludge ratio. Chemicals can be used to enhance primary treatment, but they are not often found to be cost effective.

SECONDARY TREATMENT

Following preliminary and primary treatment, the wastewater contaminants remaining consist of colloidal matter that is highly organic and a small amount of dissolved organic matter, nutrients, and dissolved inorganic solids. Colloidal and dissolved organic matter are both amenable to biological secondary treatment. There are numerous methods used to accomplish biological treatment, but most can be categorized as either suspended-growth or fixed film systems. Suspended-growth processes involve sustaining an active culture of microorganisms in suspension within a reactor, where air or oxygen is introduced to maintain biological activity. The demand for oxygen within a biological reactor can be met in various ways, but most aeration devices currently being used may be classified as either diffused, dispersed, or mechanical aeration systems.

Overall, the aeration devices used for the activated-sludge system represent the most significant consumers of energy within a wastewater treatment system. The ability of any type of equipment to dissolve oxygen in wastewater will depend on many factors, including diffuser device type, basin geometry, diffuser depth, turbulence, ambient air pressure, temperature, spacing and placement of the aeration devices, and diurnal variations in wastewater flow and organic load.

The most important energy decision for secondary treatment is made when the design is finalized. Once a system is installed, it is up to the operator to maintain the system at peak efficiency. Energy considerations for aeration systems, including the method for calculating energy use, are covered in greater depth in Chapter 8.

Return activated sludge (RAS) pumping is also required for the activated-sludge process. Return activated sludge rates are usually expressed as a percentage of the influent flow and typically range between 40 and 100% of the influent flow. Return activated sludge pumps require almost as much energy per unit as influent pumps but usually have slightly lower TDH and energy requirements. Waste activated sludge (WAS) is usually removed from the process with pumps or air lifts. While the TDH may be considerably higher for WAS pumps, flow rates are on the order of 1 to 3% of the influent flow.

Fixed film processes involve a fixed media on which microorganisms grow. The media can be any type of nonbiodegradable material that will maintain its structure when exposed to water for a long period of time. Media commonly used are plastics, rock, slag, coal, and redwood. Fixed film processes such as trickling filters typically require that the pretreated wastewater be pumped a second time in the overall WWTP process, with oxygen being supplied as the wastewater *trickles* down through the media. These units often require recirculation to provide both a minimum wetting rate as well as a more uniform wetting rate. Recirculation rates may be as much as three times the daily influent flow. Rotating biological contactors (RBCs) typically do not require repumping of the wastewater to the unit, but often use pumped recirculation. Rotating biological contactors use rotational energy for supplying oxygen rather than pumping. Fixed film processes use less energy than suspended-growth processes, but do not achieve as high a degree of treatment. Additionally, fixed film processes are more prone to odors. Once odor systems are installed, the relative energy efficiencies or at least overall operating costs of fixed film processes begin to approach those of suspended-growth processes.

Most secondary treatment systems use a secondary clarifier for capture and recirculation of the biological solids created in the process. Secondary clarifiers are not large consumers of energy and typically use fractional horsepower drives.

Energy use curves for various secondary treatment processes were developed by the U.S. Environmental Protection Agency (U.S. EPA, 1978). These curves provide energy utilization rates (in kilowatts per year) for a WWTP flow capacity range of 0.05 to 5 m^3/s (1 to 100 mgd).

DISINFECTION

Disinfection of final effluent destroys most of the microorganisms remaining in the wastewater after treatment, including pathogens. At present, the most common method of disinfection is by the addition of chlorine gas solutions or hypochlorite solutions. Because of recent concerns of carcinogenic compound formation with the use of chlorine, two alternative disinfection processes have received renewed levels of interest: ozonation and ultraviolet (UV) radiation. They require considerably greater quantities of primary energy than chlorination. The largest amount of energy use in chlorination is the secondary energy required in the manufacture of the chlorine or chlorine compound. Energy consumers for several disinfection processes are summarized as follows:

- Chlorine gas—evaporator heater, pumping of dilution water, and pumping of chlorine solution;

- Hypochlorite—dilution water pumping and metering pumps for hypochlorite;
- Ozone—air compressor, air dryer, ozone generator, and pumping of dilution water; and
- Ultraviolet—power for UV tube lighting.

The capital cost and energy required for ozonation and UV disinfection are high compared to those required for chlorine gas or hypochlorite disinfection. However, when fire codes require containment areas with exhaust ventilation through an absorbing solution for liquid chlorine and sulfur dioxide, alternate disinfection methods become attractive.

ADVANCED WASTEWATER TREATMENT

GRANULAR MEDIA FILTRATION. Filtration is used in wastewater treatment as an advanced or polishing step for removal of suspended solids and associated organic matter. The filtration operation uses energy to overcome head loss through the filter media and for backwashing. In filtration operations that use gravity flow, the design engineer has built into the design a fixed head loss. Pumps somewhere in the system must overcome the entire predesignated head loss regardless of the actual filter head loss. In totally enclosed pressure filters, head loss builds as the media becomes fouled. Although the head loss varies with the buildup of debris, overall losses may be as severe or worse than with open filters. Backwashing restores the head loss back toward the original condition. Ancillary processes such as filter-aid chemical pumping, polymer feed systems, and surface air scouring and wash systems use small amounts of energy for mixing, pumping, and air compressor power, respectively.

ACTIVATED CARBON ADSORPTION. Energy is consumed in the activated carbon adsorption process to overcome head loss in the contacting mode, which is similar to the built-in head losses with granular media filtration described above. The spent carbon requires fuel for thermal regeneration to operate a pyrolysis furnace or to furnish steam. Some carbon systems use chemicals such as caustic soda for regeneration. Electricity is required to operate the carbon transfer system.

CHEMICAL TREATMENT. Energy required for chemical treatment can be divided into four parts: chemical feed, rapid mix, flocculation, and sedimentation. Sludge pumping and chemical regeneration such as lime recalcination can consume large amounts of energy. In chemical treatment processes, the largest amount of energy use is the secondary energy required in the manufacture of the chemicals.

BIOLOGICAL NUTRIENT REMOVAL. Control of nutrients in WWTP effluents is becoming increasingly important as more stringent discharge requirements are imposed. In freshwater receiving waters, phosphorus has generally been considered the limiting nutrient in eutrophication. However, for some estuarine and other water bodies, control of both phosphorus and nitrogen may be necessary to prevent deterioration of water quality.

Within the last 20 years, significant developments have occurred in the technology for biological removal of nutrients from wastewater (WPCF, 1983). Removal of both nitrogen and phosphorus from wastewater can be reliably achieved at a life cycle cost considerably less than with conventional chemical approaches. In addition, these dual nutrient removal processes require significantly fewer chemicals, thus, lower secondary energy consumption.

Historically, removal of phosphorus from wastewater has been accomplished by chemical precipitation with metal salts. However, this approach results in high chemical costs and high sludge treatment and disposal costs. More modern techniques employ biological means for phosphorus removal. Processes are employed that use anaerobic selectors to stress organisms before they are put back in an oxidative environment. The stressed organisms then take up more phosphorus than they actually need and store the phosphorus within their cells. As the anaerobic phase is an extra step in a conventional process, the additional energy required is the energy used for mixing the tank contents (keeping the organisms in suspension) in the anaerobic phase.

Conventional nitrogen removal is accomplished through biological nitrification and denitrification. Biological conversion of ammonia to nitrate (nitrification) requires oxygen to be supplied at 4.57 times the ammonia loading, which is a significant energy-consuming process. Additionally, because nitrification only occurs with low food-to-microorganism ratios, more energy is required to satisfy endogenous respiration demands. See Chapter 8 for a more thorough discussion of aeration energy requirements.

Conversion of ammonia to nitrate is often sufficient for many discharge permits. However, in some cases, total nitrogen removal is the requisite. Nitrogen can be removed by air stripping of ammonia from effluents with pHs that have been elevated above 11 with alkali chemicals. This practice is relatively inefficient and temperature-sensitive, requires considerable amounts of chemicals (usually lime and neutralizing acid), is expensive, and adds offending ammonia to the air. An alternative process is biological denitrification following a nitrification process. Biological denitrification, however, requires an organic carbon source. If done as a separate process following secondary treatment, methanol is often used as the carbon source.

A preferred but not totally efficient method for denitrification uses an aeration system with an anoxic zone partitioned off from the aerobic portion of the system. The anoxic zone is located at the beginning of the aeration tank and uses mixers of some type rather than aerators or air diffusers. Nitrified wastewater from the rear of the aeration tank is recirculated to the anoxic zone, in which the nitrate is used as a source of oxygen, and the incoming biochemical oxygen demand is used as the carbon source. As the nitrate gives up its oxygen, it breaks down to nitrogen gas and is released to the atmosphere.

The nitrate actually gives back some of the oxygen that was used in its formation in the nitrification step. While it takes 4 600 g of oxygen per kilogram of nitrogen (4.6 lb/lb) to produce the nitrate, the nitrate only gives back 2 800 g of oxygen per kilogram of nitrogen (2.8 lb/lb) (the additional oxygen that went to forming water in the oxidation of ammonia cannot be recovered). Additional amounts of energy are required for recirculation and for mixing the contents in the anoxic zone.

To be effective, the process requires high recirculation rates. Through a mass balance calculation, and assuming 100% utilization of nitrate in the anoxic zone, it can be shown that the efficiency of denitrification is related to the recirculation ratio in the following manner:

$$\text{Maximum efficiency of denitrification} = 1 - \frac{1}{N+1} \tag{7.4}$$

Where

N = the number of recycles relative to the influent flow.

For example, if the influent flow is 0.05 m^3/s (1 mgd), the recirculation is 0.1 m^3/s (2 mgd), and the number of recirculations is two, then the maximum efficiency of denitrification is 67%. This means that 33% of the nitrate would appear in the effluent and 67% would break down in the anoxic zone, yielding its oxygen and exiting the system as nitrogen gas. To attain 90% nitrogen removal would require nine recirculations; to attain 95% removal would require 19 recirculations. When high removal rates of nitrogen are required, it does not make sense to use the high recirculation rates needed. Consequently, a separate denitrification stage following secondary treatment will be required. Addition of primary effluent to the final stage of treatment can be done, but it brings up philosophical questions relating to the purification of the wastewater. More commonly, methanol is used as the carbon source, which can become an expensive process and generate additional sludge.

SLUDGE TREATMENT AND DISPOSAL

Major sludge treatment operations include thickening, conditioning, stabilization, dewatering, drying, and combustion processes. Some of these processes and associated energy requirements are described in Chapter 9 and in other manuals (U.S. EPA, 1978 and 1979). Disposal options include landfilling, land application, and beneficial use products such as dried pellets, compost, and stabilized sludge soil.

Energy is required for pumping the sludge to or between the processes and may be required for transporting the final sludge product by truck, rail, barge, or pipeline to the final disposal site.

MISCELLANEOUS ENERGY USES

There are many auxiliary processes in a WWTP that require energy. Examples are seal water, service water, auxiliary service compressed air, instrument air, hoisting cranes, sump pumps, and sometimes potable water. Some minor energy uses include instrumentation, electric-operated valves, gas pilots for boilers, digester gas flares, unit heaters, and portable steam units. Additionally, an emergency power supply may be available in the form of engine generators.

Wastewater treatment plants have buildings for housing equipment and personnel for facility operation, maintenance, laboratory, and administration. All of these buildings use energy for space heating during colder weather in cold climates and air-conditioning during warmer weather. Energy is also used for air-handling units for ventilation, odor control units, and for lighting work areas, outside areas, and interconnecting gallery areas.

REFERENCES

U.S. Environmental Protection Agency (1978) *Energy Conservation in Municipal WastewaterTreatment.* MDC-32, Washington, D.C.

U.S. Environmental Protection Agency (1979) *Process Design Manual for Sludge Treatment and Disposal.* EPA-625-1-79, Washington, D.C.

Water Pollution Control Federation (1983) *Nutrient Control.* Manual of Practice No. FD-7, Washington, D.C.

SUGGESTED READINGS

University of Florida (1986) *Operations and Training Manual on Energy Efficiency in Water and Wastewater Treatment Plants.* Training Res. Educ. Environ. Occupations Cent., Gainesville, Fla.

U.S. Environmental Protection Agency (1985) *Handbook—Estimating Sludge Management Costs.* EPA-625/6-85, Washington, D.C.

Water Environment Federation (1992) *Design of Wastewater Treatment Plants.* Manual of Practice No. 8, Alexandria, Va.; Manual and Report on Engineering Practice 76, Am. Soc. Civ. Eng., New York, N.Y.

Chapter 8
Aeration Systems

Electricity for aeration systems typically represents a substantial portion of the total wastewater treatment plant (WWTP) electrical energy use and overall operating cost at most secondary and advanced WWTPs using diffused or mechanical aeration. For any such WWTPs, focusing on efficient aeration system energy use is, therefore, an obvious first step in energy conservation.

In determining oxygen requirements for a new activated-sludge system, engineers typically will use several rational approaches, but will also use a large safety factor to ensure that an adequate oxygen supply is always available for peak load conditions. Operators, on the other hand, typically use mixed liquor dissolved oxygen (MLDO) as their guide to increasing or decreasing oxygen supply. While this would seem to be an adequate means of control, a number of factors can affect not only the efficiency of oxygen transfer and oxygen use, but, more important, the efficiency of energy use. Commonly encountered conditions such as numbers of aeration tanks in service, air rate per diffuser, presence of nitrification when not required, food-to-microorganism (F:M) ratio, aerator submergence, and air filter cleanliness can influence overall efficiency.

Many textbooks give various empirical formulas for calculating oxygen requirements that are based on how much of the biochemical oxygen demand (BOD) removed is converted to carbon dioxide and water and how much is converted to biomass. Factors in each equation, unfortunately, are typically listed in broad ranges, requiring the operator to deduce from WWTP operation what specific factor values to use. This, together with empirical relationships linking solubilities of oxygen in clean water to that in wastewater, makes use of oxygen transfer calculations by operators appear meaningless.

DETERMINING OXYGEN REQUIREMENTS

Generally, the amount of actual oxygen required in the stabilization of raw or primary treated wastewater is attributed to the carbonaceous demand, endogenous demand, nitrogenous demand, the initial dissolved oxygen (DO) deficit of the wastewater or primary effluent, and the degree of stabilization required or achieved. The carbonaceous demand is actually a function of the secondary system loading, but is frequently determined by the actual or desired BOD removal across the system. Typically, half the BOD removal goes to respiratory byproducts (carbon dioxide and water), while the other half goes to the formation of new sludge (biomass). Typically, it is biomass carryover in the effluent that causes residual BOD rather than unoxidized organic matter.

Computing the oxygen requirements from BOD loading or removal can be complex. Approximately half of the BOD removed produces sludge rather than consuming the stoichiometric amount (chemically exact amount) of oxygen for combustion. Eckenfelder and O'Connor (1961) presented a classic formula for converting BOD removal and endogenous respiration demand to oxygen requirements for the activated-sludge process. Houck and Boone (1981) have converted this formula to one that is a function of an F:M ratio redefined in terms of removal rather than loading. Their formula can further be simplified if a standard 90% BOD removal across the aeration system is assumed. Their formula would then read

$$R = 0.75 + [0.056/(\text{F:M})] \qquad (8.1)$$

Where

R	=	pounds of oxygen required per pound of BOD_5 (g/kg) removed, and
F:M	=	conventional F:M ratio, pounds BOD_5 applied per day per pound of mixed liquor volatile suspended solids (MLVSS) (g/kg) under aeration.

The carbonaceous oxygen requirements can be computed simply by multiplying the pounds (kilograms) of BOD_5 removed per day by the value calculated for R.

This formula appears reasonable for F:M ratios greater than 0.1. For ratios less than 0.1, other factors must be considered because the equation is no longer valid, as calculated oxygen requirements rise more rapidly than those actually required.

Oxygen utilization is only a part of the BOD removal activity, with the remaining part of BOD removal going to sludge production. Using factors presented by Eckenfelder and O'Connor (1961) for production of sludge, which show that 0.5 kg (1 lb) of volatile suspended solids (VSS) requires approximately 0.64 kg (1.42 lb) of oxygen to be oxidized, one can show the proportional relationship between oxygen required per kilogram (pound) of BOD removed and the VSS produced per kilogram (pound) of BOD removed.

Nitrogenous oxygen demand is often more difficult to estimate. If biological nitrification is not required, it is uneconomical to allow it to occur. When not required, the nitrogenous demand can be considered to be negligible unless nitrite or nitrate is observed in the effluent. In the case of WWTPs required to nitrify, effluent results for nitrite and nitrate nitrogen can be summed and multiplied by the conversion factor, 4.57, to determine the total nitrogenous oxygen demand. An alternate and preferred method would be removal of ammonia across the aeration system. Either method is susceptible to error because of the complex chemistry of nitrogen in the activated-sludge process. Some ammonia is typically converted to organic nitrogen in cell mass, meaning that some of the ammonia removal does not result in oxygen use, and some nitrate may denitrify and not be present in the effluent; whereas, some organic nitrogen may be converted to ammonia and become oxidized and not be represented in the measured ammonia reduction across the process.

If nitrification occurs, the oxygen for nitrification is computed by multiplying the pounds (kilograms) of nitrogen to be oxidized per day (usually ammonia to be removed) by 4.57. An amount for increasing the DO of the aeration tank influent from zero levels to the effluent MLDO level of approximately 2 mg/L can be included but in most cases is small enough to neglect.

Care must be taken in selecting data for use in determining oxygen requirements. Inaccurate flow measurement, sample streams contaminated with recycle, non-flow-paced sampling, intermittent (1 to 3 samples per week) sampling, and the laboratory test (BOD_5) used as the industry standard may contribute to erratic data from which to try to develop meaningful relationships (Kennedy and Boe, 1985).

OXYGEN TRANSFER EFFICIENCY

Oxygen is typically supplied to activated-sludge systems using some type of aeration system, which injects air bubbles into the mixed liquor. The two major types of aeration system commonly used are diffused air and mechanical aerators. Diffused air systems have as major components blowers and diffusers. Mechanical aerators consist of an impeller or set of brushes or disks that are driven on a vertical or horizontal shaft. The primary purpose of aeration equipment is to dissolve the oxygen from the air/gas phase into the liquid to satisfy the oxygen demands within the mixed liquor. Additionally, aeration equipment is used to maintain sufficient mixing to keep solids in suspension.

Transfer of oxygen from air to water or wastewater is relatively inefficient and is energy-intensive. The fraction of oxygen in the air bubbles supplied that dissolves in the mixed liquor is the oxygen transfer efficiency (OTE). The OTE depends on the conditions of aeration and the equipment. Conditions differ between WWTPs and between aeration tanks at any WWTP, and also vary with time in any tank.

To test and compare efficiencies of various types of aeration equipment, engineers have devised standardized test procedures using clean deoxygenated water under standard conditions of 20°C and one atmospheric pressure. Differences from these conditions must include a correction. For diffusers, OTE increases with depth. Rather than standardize depth in the tests, standard tests are typically conducted at the depth for which the diffusers are intended.

Operating aeration equipment under standard test conditions yields the standard OTE. The standard OTE divided by the energy used per unit weight of oxygen delivered is the standard aeration efficiency (SAE). The SAE expresses the weight of oxygen dissolved per unit of energy used by the aeration equipment.

The actual, or field, OTE differs from standard OTE because real conditions with wastewater differ from standard test conditions. The ratio of normalized field OTE to standard OTE reflects the effects of such contaminants and is referred to as the *alpha factor.* Oxygen transfer efficiency depends on various parameters of water quality. These include the presence and concentration of surfactants (surface active agents), such as detergents, and dissolved and suspended solids. The properties of detergents that cause foaming or the formation of bubbles that resist collapsing also inhibit oxygen transfer from bubbles to water, even in dilute solutions like wastewater.

Surfactants at air bubble surfaces cause the bubbles to become more rigid and spherical. Surfactants typically depress the surface renewal rate at the air–water interface, which reduces alpha below unity for a diffused aeration system.

AERATION EQUIPMENT TYPES. Aeration equipment is widely used to deliver oxygen to various biological, physical, and chemical processes, most notably at WWTPs. The total capacity of aeration equipment in North America was approximately 1.3 million kW in 1982. The enormous energy consumption and the wide range in energy efficiency of aeration systems continue to stimulate development to improve the performance of aeration equipment.

Aeration primarily aims to dissolve oxygen into mixed liquor to satisfy the incoming oxygen demand at minimum cost. An additional requirement includes maintaining sufficient mixing to keep the aeration tank contents in suspension. Aeration is also used in aerobic sludge digestion.

As previously mentioned, aeration consumes 40 to 70% of the energy used in an activated-sludge WWTP. Effort and expense directed toward improving and maintaining aeration energy efficiency can be cost effective. Improving energy efficiency often requires a team effort by design, operation, and maintenance personnel.

Diffused Aeration. Diffused aeration delivers compressed air through porous or perforated diffusers into wastewater to dissolve oxygen. Several factors affect the energy efficiency of diffused aeration.

Deeper tanks enhance OTE by lengthening the bubbles' rise and by increasing the hydrostatic pressure on the bubbles, both of which hasten oxygen transfer. However, pressure compresses the bubbles, reducing their surface area, and deeper tanks require more energy to pressurize the air, necessitating more expensive equipment. Additionally, maintenance costs will increase dramatically as blower or compressor pressure increases. In practice, site conditions determine aeration tank depth more than energy efficiency does. Because of high excavation costs and the need for tank walls with greater structural strength as a result of higher groundwater hydrostatic pressure when the tank is empty, deep tanks are rarely used unless the site is severely constrained with limited area for build-out. Figure 8.1 plots results of clean water tests of various aeration devices at different depths. Some devices show a fairly constant energy efficiency with varying depth (Houck and Boone, 1981).

Diffusers with finer pores tend to produce higher values of standard OTE. However, reducing the pore size does not necessarily improve energy efficiency to the extent that it improves standard OTE. This is because pressure

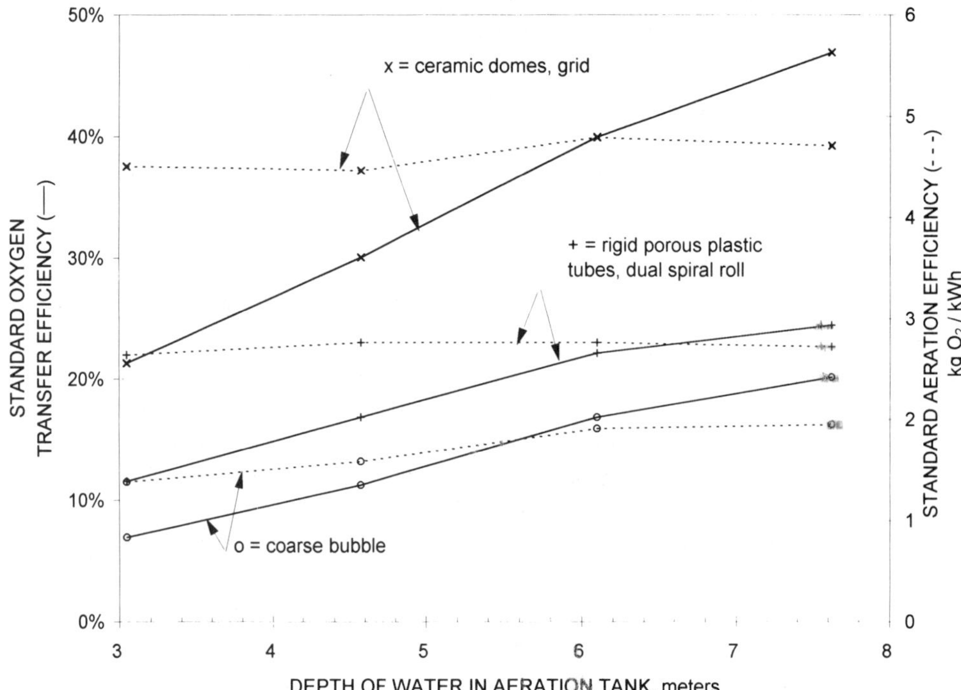

Figure 8.1 Comparative energy consumption for porous diffusers versus orifice diffusers (U.S. EPA, 1989)

differential increases across fine-pore diffusers as pore size diminishes. Additionally, pores of new or cleaned diffusers often become fouled quickly, with a concurrent decrease in OTE.

Fine-Pore System Economic Considerations. The primary reason to install fine-pore aeration diffusers is to save money through reduced energy usage. Energy savings result from the higher OTE of fine-pore diffusers. However, the higher capital costs of diffusers, blowers, and accessories must be considered, plus the higher maintenance costs associated with cleaning of porous diffusers. Another hidden cost is the loss of efficiency between cleanings. Heavily fouled fine-pore diffusers have OTEs comparable to coarse-bubble diffusers.

OPERATIONAL CONSIDERATIONS

MIXING. Determining the optimum level of power to use in an activated-sludge system can be complicated. Minimum levels of power per unit volume are provided for adequate mixing, especially at low F:M ratios, to prevent solids from settling. For most aeration tanks, 15 W/m^3 (75 hp/mil. gal) of tank capacity or more will be required to prevent settling of the mixed liquor suspended solids (MLSS). Where mixing requirements exceed oxygen supply requirements, the operator should obviously consider reducing the number of tanks and increasing the MLSS concentration, thereby retaining the same F:M ratio and improving electrical efficiency.

AIR RATE PER DIFFUSER. The air rate per diffuser is critical for all diffuser types. An insufficient air flow can cause clogging of the diffuser and solids settling in the tank. Too high of an air rate causes efficiency to drop off rapidly because small air bubbles have a greater tendency to collide and form larger bubbles. Additionally, a higher air rate corresponds to greater energy input, which imparts increasing vertical velocity currents and simultaneous faster-rising bubbles to the liquid. Fine-pore diffusers generally have the narrowest range of acceptable air rates and typically have their greatest oxygen transfer per megajoule (horsepower-hour) at their lowest acceptable air flow. From the lowest acceptable air rate per diffuser to the highest, OTE may decrease by as much as 25%.

With porous diffusers, OTE tends to decline with increasing air flow rate, but for orifice (coarse-bubble) diffusers, OTE is substantially independent of air flow over their rated range, though their OTEs are less than for porous diffusers (U.S. EPA, 1989). A corollary is that field OTE for porous diffusers tends to improve with increasing diffuser surface area. Limited studies confirm this for ceramic diffusers at a fixed air-flow rate. That is, densely packed and lightly loaded diffusers tend to be more energy efficient. For a fine-pore system, one should consider installing the maximum number of diffusers and operating all aeration basins while maintaining the air flow at or near constant optimum flow. However, the practicality of densely placing diffusers is generally limited in practice by the additional construction cost, reduced access for cleaning, and real-life varying loading conditions.

ALPHA. One of the most important and least understood concepts in oxygen transfer relates to the alpha factor. This is a factor that compensates for the effect of surfactants and other wastewater constituents that restrict the transfer of oxygen in wastewater. Literally, it is the ratio of the oxygen transfer rate in wastewater to that in clean water. A greater problem exists in fine-pore diffusion systems, in which high rates of oxygen transfer from air to water are expected, than in coarse-bubble or mechanically aerated systems. To some degree, the operator may be able to improve the alpha factor. It is known that alpha increases from the inlet of the aeration tank to the outlet as the offending contaminants are biodegraded. This was dramatically shown in a study in Whittier Narrows, California (U.S. EPA, 1989). The alpha factor is most pronounced in fine-pore systems, in which the alpha may be 0.4 or less at the inlet and increase to 0.8 or more at the outlet. Consequently, the operator can improve the average alpha by increasing the MLSS and operating at a lower

F:M ratio to more rapidly remove the surfactants. Before doing so, however, the operator must consider conflicting factors that will increase oxygen requirements, such as higher endogenous respiration demand and nitrification demand when not required. The operator must determine whether the resulting increase in alpha by operating at a lower F:M is more cost effective and must also be concerned with development or aggravation of *Nocardia* foam or problems from other types of filamentous bacteria.

Waste Loading Distribution. Among activated-sludge process modifications, some are defined by the distribution pattern of waste loading to the aeration tank. For example, plug-flow tanks receive inflow at the upstream end, while step-feed tanks distribute the wastewater feed along the length of the tank. Complete-mix activated sludge produces a uniform waste loading throughout each aeration tank.

The degree of treatment increases uniformly along the flow path in a plug-flow aeration tank and in stages in a step-feed tank, but it remains uniform throughout complete-mix aeration.

Field OTE improves with the degree of treatment; that is, efficiency of field OTE increases toward the standard OTE. Field OTE has been observed to increase along a plug-flow aeration tank, paralleling the improved degree of treatment.

Step-Feed/Complete-Mix. Operating in the step-feed or complete-mix modes may also result in a lower overall alpha by distributing surfactants more uniformly throughout the aeration tank and allowing for their dilution and more rapid removal by adsorption. A study done in Madison, Wisconsin, demonstrates the benefit of step feeding on improving alpha (U.S. EPA, 1989). Another possible side benefit of operating in either of these modes for a fine-pore system is the probable reduction of diffuser sliming, which is known to adversely affect the inlet area diffusers in plug-flow systems. Sliming will increase air system backpressure and reduce diffuser transfer efficiency, both of which contribute to overall system inefficiency. Sliming may also induce coarse bubbling as sections of diffuser become clogged and air is forced through remaining open areas.

NITRIFICATION. The occurrence of nitrification that is not required is probably the largest aeration system inefficiency. For each pound (kilogram) of ammonia nitrogen oxidized to nitrate, 4.57 lb (4 570 g) of oxygen is consumed. Additionally, when nitrification occurs, it is typically associated with a lower F:M ratio, typically on the order of 0.1 lb BOD/day/lb MLVSS (1.2 mg/kg·s) under aeration or less. Using Equation 8.1 for *R* value, one would find that the *R* value for a conventional aeration F:M of 0.3 would be 0.94 lb O_2/lb BOD (940 g/kg) applied, whereas the *R* value for a nitrifying system with F:M at 0.1 would be 1.31 lb O_2/lb BOD (1 310 g/kg) applied. As an example of the effect on oxygen requirements of nitrification, assume that 0.1 lb of ammonia-nitrogen is oxidized per pound of BOD (100 g/kg) (that is, NH_3-N concentration at 10% of the BOD concentration). For full nitrification, then, 0.46 lb O_2 (0.21 kg) would be required for the ammonia-nitrogen oxidation. Based on the lower F:M required for nitrification, a higher *R* value develops, so that 1.31 lb O_2 is required per lb of BOD (1 310 g/kg) applied. The total nitrifying system requirement relative to the influent BOD loading to the aeration system is 1.77 lb O_2/lb BOD (1 770 g/kg) applied. Compared with the conventional system loading with 0.3 F:M and an *R* value of 0.94 lb O_2/lb BOD (940 g/kg) applied (see Table 8.1), there is an oxygen requirement increase of 88%—nearly double. Because approximately 40% of the extra oxygen goes to sludge oxidation through endogenous respiration, it may not have a significant effect on overall energy usage if the waste sludge is to be subsequently stabilized through aerobic digestion. However, it may be of concern if any other type of processing is used. If anaerobic digestion is used for stabilization, there will be a corresponding net reduction in methane produced.

Table 8.1 Effect of nitrification on oxygen requirements

Nitrification	F:M[a]	Carbonaceous demand, lb O_2/lb[b] BOD[c]	Nitrogenous demand, lb O_2/lb BOD	Total demand, lb O_2/lb BOD
Required	0.1	1.31	0.46	1.77
Not required	0.3	0.94	0.00	0.94

[a] Food-to-microorganism ratio.
[b] lb/lb × 1 000 = g/kg.
[c] Biochemical oxygen demand.

Partial nitrification is difficult to prevent during summer in northern climates and practically year-round in southern or warmer climates. Typically, an effort is made to operate at the highest F:M that will obtain the necessary BOD removal while not allowing MLDO levels to exceed 2 mg/L at any time. Better control of the MLDO throughout the aeration system would be more effective, but few systems are equipped with the necessary instrumentation and valving. Periodic analysis of effluent for presence of total oxidized nitrogen should be performed in all WWTPs not required to nitrify, especially when wastewater temperatures exceed 20°C. Wastewater at a significantly greater temperature is easily induced to nitrify and requires extra attention.

MIXED LIQUOR DISSOLVED OXYGEN. The most common cause of aeration system inefficiency is excessive MLDO. The reason for this is not as self-evident as it might appear. In addition to supporting and enhancing nitrification as previously discussed, high DO levels are generated at the expense of high energy usage. The closer the MLDO is to saturation, the greater the resistance for dissolution of oxygen will be—hence, the lower the OTE.

Saturation values for oxygen in dissolved air systems are approximately 7.0 mg/L, but are even lower at temperatures greater than 25°C.

Using standard oxygen requirement calculations, it can be seen that a change as small as 1 mg/L in MLDO can result in a system inefficiency of 15%.

In WWTPs with automated DO control, inefficiency caused by excessive MLDO should not be a serious problem as long as the system is set properly and well maintained, and provides reliable results. However, it typically is a problem when manual control is practiced. Operators will tend to overcompensate with manual systems to "keep ahead of the game" should an increase in load occur before the next MLDO reading. In medium to large WWTPs, the payback period for installing automated blower control from MLDO probes can be quite short.

A recommend energy conservation measure (ECM) is as follows:

- Use of DO probes in aeration tanks with blower feedback control systems can result in substantial energy and cost savings with quick payback. Their installation is highly recommended.

Figure 8.2 illustrates the potential for improving aeration energy efficiency. Solid lines in Figure 8.2 show temporal profiles of DO over a 24-hour test period in two parallel, completely mixed aeration trains. Lines labeled *manual* and *automatic* represent oxygen concentrations in aeration tanks with manual and automatic control of DO, respectively. The upper DO concentration is generally higher than the lower line. This experience suggests how automatic controls can significantly improve the match of the concentration of DO to the target level. In this example, the estimated energy savings with automatic controls based on the relative air flows for oxygenation averaged 33%. Actual energy savings realized with automatic controls were higher at 38%, presumably because both air flow and blower pressure declined.

Actually, while beneficial for the reasons noted, most automated systems can be improved. Most automated systems are triggered by MLDO readings from the tail end of the aeration tank. In plug-flow systems, the DO sag caused by a peak loading would not be observed at the end of the tank until well after the peak has entered the tank. Consequently, during load changes the system gets out of balance. How serious a problem this is depends on hydraulic retention times and frequency of load changes. Moving the DO probe back from the end by approximately one-third of the tank length may be of some benefit. Using two or more probes per tank can lead to greater control, but the costs associated with the probes, control system, and maintenance reach diminishing returns. Multiple probes may only prove of value in larger systems.

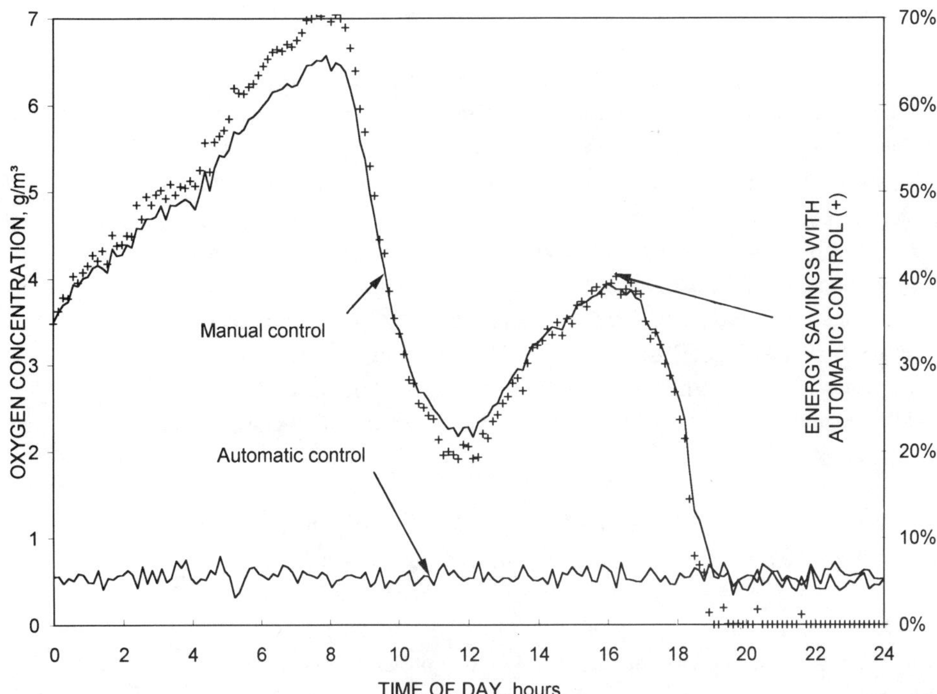

Figure 8.2 Comparative dissolved oxygen profiles for automated control versus manual operation (U.S. EPA, 1989)

Most WWTPs have characteristic daily, weekly, and seasonal loadings. With the appropriate analysis and study and some around-the-clock sampling and analysis, the wastewater can be characterized. Using this characterization study, an operator can predict increases or decreases in BOD load and can compensate for this change by making the appropriate blower adjustment in anticipation of the changes rather than after the fact. This approach can dramatically improve both treatment performance and aeration efficiency. It will help improve energy efficiency for either manually controlled or automatically controlled systems.

A recommended ECM is the following:

- Characterize the waste stream to the aeration system to anticipate loading changes by conducting around-the-clock sampling studies.

PROCESS MONITORING. A good process monitoring program is essential to energy efficiency of aeration systems. Thus, increasing blower energy use or rising air pressure may reflect the onset of fouling. Aeration tank surface patterns can reflect the distribution of air and the size of air bubbles, which are significant to aeration efficiency. A continuing operation and maintenance (O & M) effort is needed for cost-efficient aeration. Constant process testing, operating plan reevaluation, and careful implementation of operating strategies are needed. This is a cyclic process.

Ideally, air delivery is controlled to match the instantaneous oxygen demand at representative points in the aeration tank. This requires monitoring DO concentrations at selected points in the mixed liquor. Operation involves continual adjustment of the air delivery so that oxygen concentrations remain near their target level.

The target oxygen concentration is the minimum level consistent with satisfactory operation of the activated-sludge process. It is determined by test and by experience for a particular WWTP to sustain the oxygen requirements of the activated-sludge flora. For activated-sludge processes, the target level is typically near 1.5 to 3 mg/L if nitrification is practiced. Automated process controls can generally match air delivery to oxygen demand more closely than manual controls.

FOULING OF POROUS DIFFUSERS. Fouling of porous diffusers can originate from the air side, water side, or both sides, and can be manifested as pore fouling or surface fouling. Pore fouling constricts pores. Surface fouling, conversely, produces the effect of enlarged effective pore sizes. As surface fouling increases, bubble size increases and coarse bubbling may ensue.

Wastewater foulants on porous diffusers include

- Solids and salts in mixed liquor that may penetrate diffusers;
- Biological growths (biofouling), mineral deposits, or fine sand on the outer surfaces of diffusers; and
- Solids from improperly filtered air or from flaking of internal air line surfaces on the underside.

Biofouling is often the result of high organic loading (that is, high F:M) whereby certain organisms attach themselves to the porous media close to the oxygen supply and feed on the soluble organics in the wastewater. It is often found in greater abundance at the entry area of plug-flow tanks. It also develops from high-strength industrial dumps that cause short-term high F:M ratios. This type of fouling can often be rectified by isolating the tank and allowing aeration to continue to burn off the fouling organisms. Once the food is gone, the organisms undergo endogenous respiration, decline in size, and eventually slough off the diffusers.

Air foulants include particulates from improperly filtered air and other material originating after filtration. Improved air filtration and corrosion-resistant air pipelines have made air-side fouling of porous diffusers uncommon. Some postfiltration foulants to consider include

- Diffuser system degradation products such as rust or flaked paint,
- Mixed liquor solids leaked into the air pipes, and

- Construction debris remaining from before initial startup of the air delivery system or entering on subsequent reopening.

Fouling relates specifically to field OTE and expresses the field OTE as a fraction of its value when the diffusers were new. Thus, fouling expresses any progressive and significant decline in field OTE.

Evaluation of diffusers for fouling is particularly critical after an operational upset. For example, a loss of air to the diffusers (especially, ceramic diffusers) could result in intrusion of mixed liquor solids, causing the diffuser pores to become fouled. Loss of air to the diffusers may necessitate cleaning all diffusers. Cleaning diffusers after a power failure is seldom a trivial task. Reliable power and a standby supply can be cost-effective precautions against costly power failures. Many newer aeration devices include check valves close to the diffuser and small air plenums that limit the amount of mixed liquor that can penetrate the diffusers on loss of pressure.

DIFFUSER CLEANING. Air-Side Fouling. After construction or any opening of the air distribution system, thorough purging or cleaning of the system is essential. Air filters must be maintained to work properly and to prevent them from creating air-side fouling. Poorly maintained viscous impingement air filters or air compressors may foul diffusers.

Liquid-Side Fouling. Diffuser cleaning is largely or entirely directed toward the water face of the diffusers. Cleaning involves dewatering aeration tanks, which exposes both personnel and the facility to certain perils. Personnel safety is paramount, but the risk of damage to the diffuser system must not be overlooked. Polyvinyl chloride (PVC) pipework and components are particularly susceptible to damage. Unforeseen and excessive thermal expansion and contraction while the tank is dewatered can also damage PVC components.

Also, PVC can degrade under the ultraviolet component of sunlight. Polyvinyl chloride pipework and fittings should be covered with water even when not in use. Air must be blown continuously through water-covered diffusers. Emptying tanks during extremes of temperature should be avoided. Ice rafts may damage equipment. Falling influent may damage pipework or diffusers.

Cleaning of Fouled Diffusers. Several cleaning methods exist for ceramic fine-pore diffusers. For *ex situ* cleaning, the diffusers are removed from the aeration tank. For *in situ* cleaning, the diffusers are cleaned in the tank. *In situ* methods are available for either drained or full tanks.

Ex situ methods include refiring of diffusers, high-pressure water jetting, and washing with silicate-phosphorus, alkali, acid, or detergent. Refiring the diffusers in a kiln destroys organic foulants but may incorporate foulant residues in the porous matrix. *Ex situ* methods are expensive and seldom used.

In situ methods include physical, chemical, and biological procedures. The diffusers may be physically washed with water, air, or steam. Other physical procedures may include sandblasting, ultrasonic vibration, flaming, drying, or air bumping. Chemical procedures may include addition of gaseous compounds such as hydrogen chloride, chlorine, or gaseous biocides to the air side or may include liquid acid or detergent cleaning of diffuser surfaces. Biological treatment involves operation of the activated-sludge process under endogenous respiration by withholding inflow.

Hosing or steam cleaning dislodges loose external biological growths and deposits. Hydrogen chloride 14% solution sprayed on diffusers accompanied by hosing before and after removes both organic and inorganic foulants. Bumping involves briefly increasing the air flow through a diffuser to the maximum rating. Bumping may be accomplished while blowers are being rotated in service, before the retiring blower is switched off. However, it is prudent to check that the additional power does not incur an increased electrical demand charge.

Acid gas injection involves the injection of hydrogen chloride or formic acid gas into the air delivered to diffusers. The air flow is held near the maximum flow rating of the diffusers for uniform cleaning of their pores. Safety training and precautions are essential. Gas injection typically continues until the pressure differential across the diffusers has stabilized, and the pressure stabilizes after approximately 30 minutes of acid gas flow.

The hydrogen chloride cleaning process is patented. It produces a hydrochloric acid solution of approximately 28% concentration in the diffuser pores. The liquid acid dissolves some inorganic salts deposited in the pores and may help remove biofilms. However, the acid does not remove dust from the air side or silicic deposits from the wetted side.

For cleaning rigid, porous, plastic diffusers, the options include all those for ceramic diffusers except refiring, flaming, and sandblasting. For cleaning perforated membrane diffusers only, *in situ* methods are generally feasible. Removal and replacement of these diffusers is excessively labor intensive. Brush scrubbing preceded and followed by hosing has been used to clean membrane tube diffusers.

Some manufacturers of perforated membrane diffusers suggest weekly or monthly air bumping, also called *flexing*. First, the air flow is stopped to collapse the diffusers onto their frames. Next, the flow is increased to two or three times normal, but within the maximum rating of the diffuser. Finally, the air flow is restored to normal. However, experience with such procedures has not always been successful.

Choice of method of cleaning is often determined by trial and error. Simple procedures such as hose washing are often tried first, with more difficult procedures following if previous cleaning was unsuccessful or short-lived. When cleaning is short-lived, operators must resort to removal of diffusers for testing of various cleaning agents or physical procedures in the laboratory.

DIFFUSED AERATION CASE HISTORIES

The following simplified summaries are restricted largely to the diffused air system alone. Generally, these case histories document improved OTE with fine-pore aeration. In many cases, a cost advantage of fine-pore aeration is evident. In several cases, preliminary simple payback periods are computed. The payback period is the initial cost associated with fine-pore aeration divided by the annual savings obtained.

FRANKENMUTH, MICHIGAN. At this 0.8-m^3/s (18-mgd) WWTP, a grid of ceramic disc diffusers replaced a stainless steel coarse-bubble aeration system. The total cost to install 2 400 diffusers was $190 000 for equipment, engineering, and installation, including 800 WWTP personnel labor-hours.

Typically, the diffusers were cleaned monthly during the first 2 years of operation and bimonthly in the third year. Cleaning saved aeration energy by reducing the operating pressure at the blower discharge, though no direct field OTE benefit was discerned. Energy usage was 0.006 6 MJ/kg (0.83 kWh/lb) of BOD_5 removed for coarse-bubble aeration, and 0.004 7, 0.005 4, and 0.007 0 MJ/kg (0.59, 0.68, and 0.88 kWh/lb) for the first, second, and third years of operation with fine-pore aeration. Energy use rose in the third year because the operators chose to deliver more air per kilogram (pound) of influent BOD_5 than before.

GLASTONBURY, CONNECTICUT. At this 0.2-m^3/s (3.6-mgd) WWTP, retrofitting 320 rigid porous plastic diffuser tubes and an air blower at a total cost of $28 000 improved field OTE to 6.5 to 7%, compared to 4 to 4.5% estimated for the original coarse-bubble diffusers.

Based on the observed blower energy reduction, the estimated annual electrical cost savings resulting from the new system were $15 000 to $18 000. This yields a simple payback period of approximately 2 years, not taking cleaning costs into account.

GREEN BAY, WISCONSIN. A 2.3-m^3/s (52.5-mgd) WWTP, this facility treats mixed municipal wastewater, seasonal vegetable canning waste, and pulp and paper mill waste. Fine-pore diffusers replaced sparged turbine aerators in some of the contact stabilization and reaeration basins.

Ceramic disc diffusers were installed in two basins, and perforated membrane tube diffusers in another two. Standard OTE test results averaged 35% and 33% for the ceramic disc and perforated membrane tube diffusers, respectively.

Operating experience showed that, over successive operating periods of approximately 6 months, the field OTEs of the fine-pore diffusers declined by a few percent from initial values of 15 to 18%. Ceramic diffusers were cleaned by hosing, hydrogen chloride gas injection, and rehosing. Membrane diffusers were cleaned by hosing, brushing, and rehosing. Cleaning restored all of the field OTE of the ceramic disc diffusers but was less effective for the perforated membrane tube diffusers, as Figure 8.3 shows.

For ceramic disc diffusers, hosing alone restored the original standard OTE of contact basin diffusers. Acid gas treatment was more effective for reaeration basin diffusers. For perforated membrane tube diffusers, hosing–brushing–hosing did clean the diffusers, but it did not completely restore the original standard OTE. Flexing the diffusers by varying the air flow was ineffective for cleaning.

Economic analysis showed a simple payback period of approximately 4.5 years for the ceramic disc diffusers, though perforated membrane tube diffusers proved uneconomic as a result of their declining field OTEs.

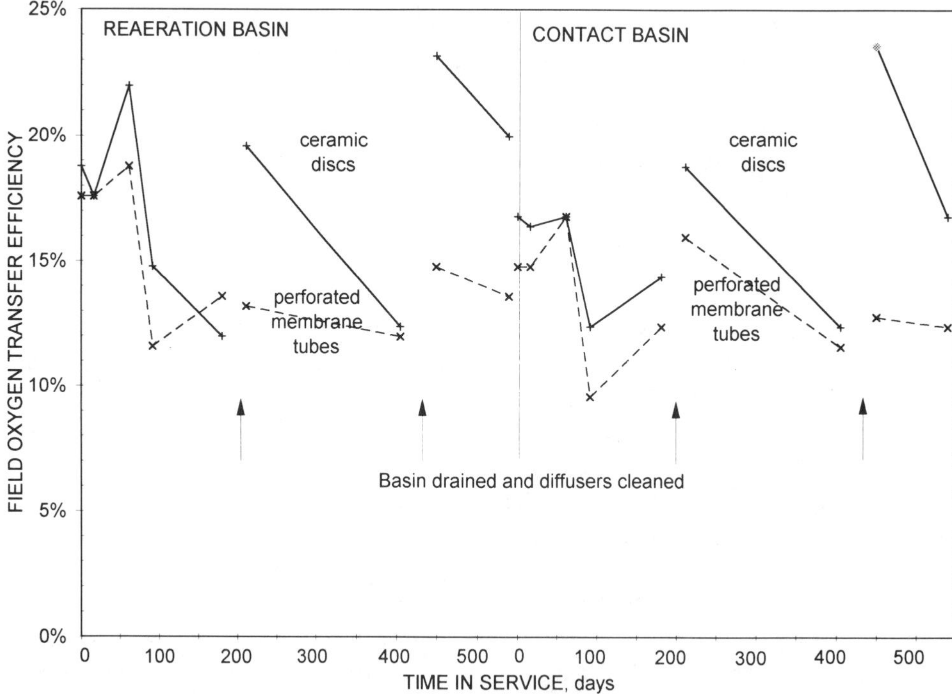

Figure 8.3 **Field oxygen transfer efficiency of ceramic disc diffusers and perforated membrane tube diffusers at Green Bay, Wisconsin (U.S. EPA, 1989)**

HARTFORD, CONNECTICUT. At this 2.6-m^3/s (60-mgd) WWTP, installing a fine-pore dome diffuser system improved the average field OTE to 10.0% (Figure 8.4) from 4.4% estimated for the original coarse-bubble spiral roll system. Offgas tests of field OTE were performed before and after cleaning the diffusers.

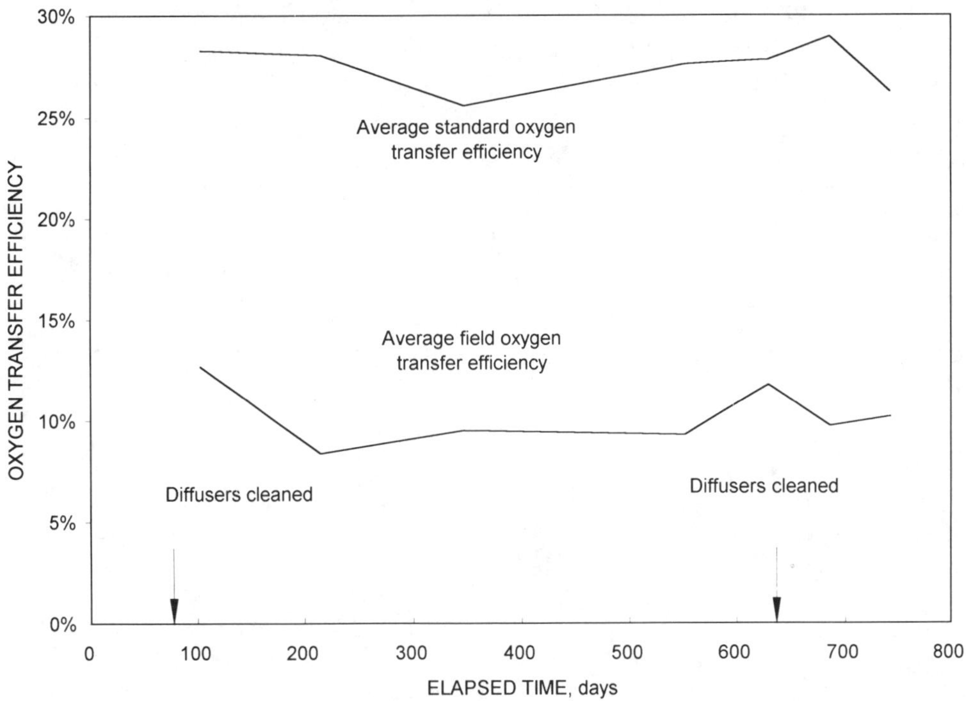

Figure 8.4 **Standard and field oxygen transfer efficiency, tank 2, Hartford, Connecticut (U.S. EPA, 1989)**

The diffusers were cleaned by hosing, spraying with a cleaner containing 22% hydrochloric acid, soaking, brushing, and rehosing. The cleaning may have had little effect on OTE but is likely to have reduced the pressure drop through the diffusers, which would have saved energy. The installed cost was less than $600 000, and annual operating savings exceeded $200 000 for the first year. This corresponds to a simple payback period of less than 3 years.

JONES ISLAND, MILWAUKEE, WISCONSIN. This 8.8-m^3/s (200-mgd) WWTP, commissioned in 1925, had plate diffusers of fused silica arranged in a ridge-and-furrow configuration. Subsequent additions used porous plates of fused alumina and silica.

After cleaning the diffusers in 1964, the field OTE was 7.6% for a spiral flow tank with tubes, 20.9% for a ridge-and-furrow WWTP with plates, and 13.8% and 15.8% for tanks with transverse and longitudinal plates, respectively. Cleaning methods included hosing, sandblasting, and acid washing. More recent cleaning (as described above for Hartford, Conn.) gave field OTE values of 17% in one tank and 19% in another.

NINE SPRINGS, MADISON, WISCONSIN. This is a 1.7-m^3/s (38-mgd) WWTP. Diffused aeration using fine-pore domes and discs was installed during the period 1977 through 1986. Extensive studies of field OTE were made. Figure 8.5 shows field OTEs in two tanks with fine-pore diffused aeration over 2 years of operation. No decline in field OTE with time is perceptible.

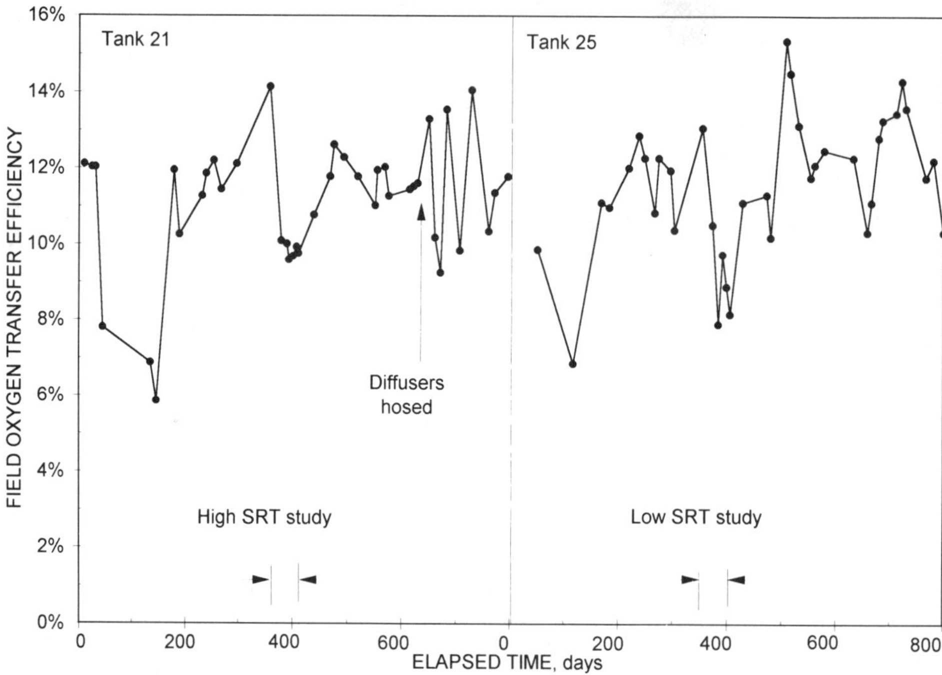

Figure 8.5 **Field oxygen transfer efficiency in two tanks, Madison, Wisconsin (U.S. EPA, 1989)**

Fine-pore diffuser aeration at this WWTP was considered cost effective, especially because cleaning of diffusers is rarely needed in the lightly loaded nitrification facility. However, the minimum air flow that the blower could deliver exceeded the oxygen demand, precluding attainment of the full economic benefit of fine-pore diffused aeration. This problem was solved in part by taking several aeration tanks out of service and diverting waste to the remaining tanks. No economic analysis was performed for this site.

RIDGEWOOD, NEW JERSEY. At this 0.1-m^3/s (3-mgd) WWTP, a dome fine-pore aeration system replaced a coarse-bubble aeration system, increasing the average field OTE from 4.8% to 9.5%. The dome diffusers were cleaned by brushing with 10% hydrochloric acid or by hosing, though changing wastewater characteristics obscured the effect of cleaning on field OTE. However, in one instance, the field OTE improved from 9.1% to 9.9% after hosing, then improved again to 11.5% after acid cleaning.

Economic analysis of fine-pore diffusion at this site was based on blower energy consumption before and after installation of the dome diffusers. Before the domes were installed, two blowers operated full time; with the domes, the blower usage fell by some 30%, as Figure 8.6 shows. The associated simple payback period was approximately 10 to 11 years considering the increased maintenance of the dome diffusers.

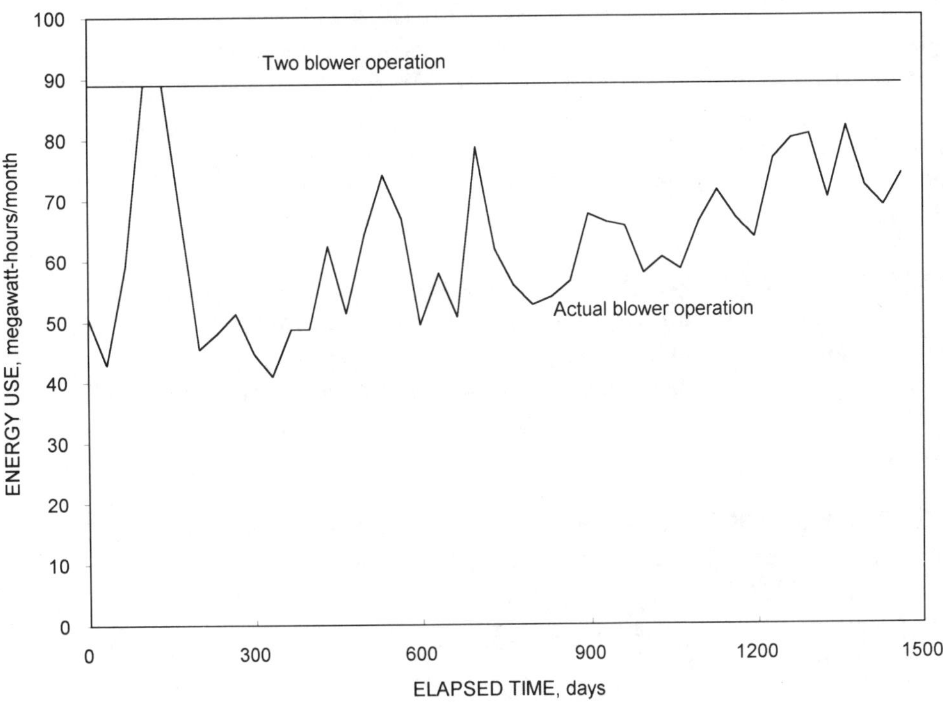

Figure 8.6 Energy savings attributed to dome diffuser aeration system, Ridgewood, New Jersey (U.S. EPA, 1989)

WHITTIER NARROWS, CALIFORNIA. Ceramic disc and ceramic dome diffusers installed in aeration tanks at this 0.7-m^3/s (15-mgd) WWTP were cleaned periodically with hydrogen chloride gas. The experimental design to evaluate cleaning procedures called for cleaning the inlet grids of diffusers every 3 months, the intermediate grids every 6 months, and the outlet grids every 9 months. Periodically, the diffusers were removed and evaluated for dynamic wet pressure differential (DWP) (WPCF, 1988), foulant mass and composition, and air distribution profiles. Test results showed that DWP could be controlled by gas cleaning, but after cleaning, DWP typically was still approximately twice the value for a new diffuser.

Offgas testing of the aeration tanks apparently showed that the field OTE improved with gas cleaning for disc diffusers but not for dome diffusers. Initial field OTEs were 10% for discs and domes, and during operation averaged 9.8% for discs and 7% for domes. These field OTEs approximately doubled the values for the coarse-bubble diffusers used before.

Based on energy savings in the high-energy-cost Los Angeles area, the simple payback period was computed as 2.8 years. Costs for cleaning diffusers had relatively little effect on overall economics at that location.

CLEVELAND, WISCONSIN. This is a 0.009-m^3/s (0.2-mgd) facility. Porous plastic plate fine-pore diffusers replaced coarse-bubble diffusers in an 11-m (36-ft) diameter package plant. The cost of $11 500 included new blowers. Annual savings in electricity reported as $2 400 equates to a simple payback period of 5 years. Periodic inspection of the diffusers has revealed only a slight growth on 5% of the diffuser surfaces, which, if desired, could be removed with 10% sulfuric acid.

PLYMOUTH, WISCONSIN. Surface aerators were replaced with ceramic disc diffusers in this 0.07-m^3/s (1.65-mgd) WWTP at a total cost of $220 000, including blowers and a blower building, and removal of the old aerators. Annual energy savings amounted to $20 000, yielding a simple payback period of 11 years. An *in situ* system for hydrogen chloride gas cleaning was installed.

RENTON, WASHINGTON. At this 3.2-m^3/s (72-mgd) WWTP, coarse-bubble diffusers were replaced with perforated flexible membrane tube diffusers at a cost of $380 000. The estimated annual energy saving is $92 500, corresponding to a simple payback period of approximately 4 years. The estimated field OTEs averaged 7.2%.

RIPON, WISCONSIN. At this 0.09-m^3/s (2-mgd) WWTP, ceramic disc diffusers replaced coarse-bubble diffusers. The estimated annual energy saving is $29 300. *In situ* hydrogen chloride gas cleaning was provided for preventive use, expected to be once per year or less.

SAUKVILLE, WISCONSIN. At this 0.02-m^3/s (0.5-mgd) WWTP, ceramic disc diffusers replaced coarse-bubble diffusers at a cost of $44 000. With annual energy savings of $7 300, the simple payback period was 6 years.

However, DWP monitoring revealed rapid and heavy fouling of the diffusers. *In situ* hydrogen chloride gas cleaning, initially expected to be needed three to four times annually, was required four times in the first 7 months of operation. The diffusers were cleaned when DWP was high, coarse bubbling occurred, or the prescribed minimum DO level could not be maintained.

COMPRESSORS AND BLOWERS

A blower is a single- or multiple-stage mechanical device used to produce relatively large volumes of air or gas at relatively low pressure, typically close to atmospheric pressure. A compressor can be classified as a single- or multiple-stage device designed to produce lower volumes of air at higher pressure. Blowers and compressors are often used in WWTPs to provide oxygen and mixing for biological treatment, solids digestion, and combustion. Blowers are more commonly used in WWTPs because there is seldom need for pressures in excess of 70 000 Pa (10 psi).

POSITIVE DISPLACEMENT COMPRESSORS. Types of positive displacement compressors include reciprocating, sliding vane rotary, liquid piston rotary, rotary lobe, and rotary helical screw compressors. They work by repeatedly trapping and compressing a fixed volume of air and releasing the compressed air at a pressure higher than the intake pressure.

Reciprocating compressors are capable of producing high pressures at low to moderate discharge rates. They use one or more pistons, each piston sliding in a cylinder, with multiple pistons arranged in parallel or in series.

Sliding vane rotary compressors are available for discharge pressures of 35 000 to 350 000 Pa gauge (5 to 50 psig) for single-stage machines or 400 000 to 850 000 Pa gauge (60 to 125 psig) for dual stage. Discharges range from 0.01 to 0.1 m^3/s (30 to 300 scfm) for single-stage, or 0.05 to 0.8 m^3/s (100 to 1 800 scfm) for dual-stage machines. They operate by compression of air trapped in pockets between rotating axial vanes and the pump casing.

Liquid piston rotary compressors are available for discharge pressures of 35 000 to 550 000 Pa gauge (5 to 80 psig) with discharges up to 7.6 m^3/s (16 000 scfm). Single-stage machines have been used as vacuum pumps up to 88 kPa (26 in. Hg). These compressors operate by using a rotating ring of liquid to compress a trapped pocket of air. The contact between air and liquid cools the air and saturates it with the liquid, a possible disadvantage for aeration.

Rotary lobe compressors use two figure-eight lobes rotating in opposite directions within a long cylindrical casing to produce low-to-medium air flows and pressures.

Rotary helical screw compressors operate similarly, with a helically lobed driving rotor and a helically grooved secondary rotor.

Centrifugal Blowers. Centrifugal blowers, otherwise known as *dynamic compressors,* work by accelerating the gas to impart kinetic energy from the rotating machinery, then converting the kinetic energy to pressure energy in the outer casing of the compressor. The two major types are axial flow and radial flow. Axial flow compressors can handle large flows of air, increasing the pressure by a factor of up to seven. The air flows essentially along the line of the rotating axle fitted with inclined vanes. Radial flow compressors can deliver higher pressures, particularly if units are arranged in series for multiple stages. The air flow enters along the line of the rotating axle, then diverts radially outward toward the casing, which discharges normally to the axle.

MECHANICAL AERATION

Mechanical aerators dissolve oxygen by thrashing the water surface to drive in air bubbles. Means of adjusting oxygenation capacity to match the oxygen demand include submergence adjustment, speed adjustment, and on–off operation.

SUBMERGENCE ADJUSTMENT. The immersion of mechanical aerator impellers can be adjusted to help match the aeration capacity to the oxygen demand. Figure 8.7 presents the results of Russian tests of the energy efficiency impacts of adjusting submergence (WPCF, 1988). These results indicate the sensitive hydraulics of the mechanical aerators tested. Operation significantly less than the optimum condition can impair energy efficiency.

Mechanical aerators create a circular hydraulic jump around the rotating impeller and form a standing wave in the mixed liquor that entraps air bubbles. Limited data from the Russian tests indicate that an efficiently operating mechanical aerator may form the hydraulic jump close to the outer edge of the aerator shroud. Adjusting the impeller submergence moves the hydraulic jump radially. Energy efficiency deteriorates as the jump is moved from its optimum location.

For floating aerators, immersion is adjusted by adding or removing weights. Cables or other means are needed to stabilize the aerator. For fixed aerators, provision is needed to control the liquid level. Given appropriate controls and equipment, the aerator immersion can be adjusted energy-efficiently to match oxygenation capacity with oxygen demand much as is done by throttling blowers.

SPEED ADJUSTMENT. Aerator power and oxygenation capacity are a function of impeller speed. Frequently, fixed mechanical aerators are equipped with multispeed motors, which provide the operator with oxygenation flexibility.

Because available aerator characteristics suggest that energy efficiency may be relatively constant over a range of operating speeds, speed variation may have an energy-efficiency advantage over submergence adjustment. The

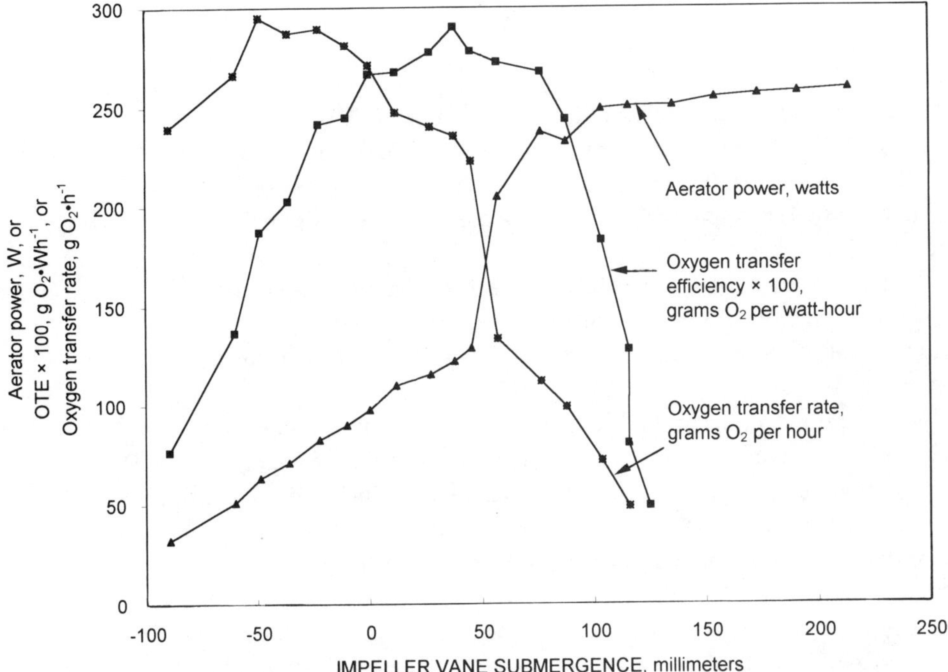

Figure 8.7 **Mechanical aerator power, oxygenation efficiency, and oxygenation rate versus impeller submergence**

feasibility of speed variation versus submergence adjustment to control oxygenation is largely determined in the initial design of the WWTPs.

ON-TIME ADJUSTMENT. On–off switching of mechanical aerators may be the most practical way to adjust oxygenation capacity to oxygen demand, but it has pitfalls. Switching must not be so frequent that it would result in overload to the electrical equipment or controls. Nor should the aerator remain off for too long. Switchings should be desynchronized to control peak demand.

Excessive off time may jeopardize the activated-sludge process. A level of DO in the mixed liquor significantly below the target level may stress the microorganisms, or the MLSS may settle. In general, at least 35 W/m^3 (75 hp/mil. gal) of energy must be provided to keep mixed liquor solids in suspension. Because of difficulty with even spacing of aerators in an aeration tank, a greater watts–to–cubic-metres (horsepower–to–million-gallons) ratio may be required to ensure adequate mixing in all reaches of the tank.

Another option that may be easy to implement is to provide or retrofit jet aerators to supplement mechanical impellers. Jet aerators may be better suited to switching than entire aerators. However, such an option still requires proper design and operation. Process control considerations include the stability of the entire aeration system. Loading fluctuations, oxygen demand estimates, and the oxygenation capacity of each aerator all must be taken into account.

MECHANICAL AERATOR MAINTENANCE AND TROUBLE-SHOOTING. Mechanical aerators require less O & M attention than fine-pore diffuser aerators, though porous diffusers are typically more energy efficient. Nevertheless, significant maintenance considerations arise with mechanical aerators.

Surface or structural damage to the impellers or shrouds will impair the original energy efficiency. Prompt identification and repair of damage and removal of entrapped material such as rope will help maintain energy efficiency.

Because some mechanical aerators are inseparable from their motors, scheduled lubrication and preventive maintenance of motors are essential. Regular maintenance according to manufacturer's recommendations is critical to energy efficiency and reliable operation. Personnel should note noise, vibration, and temperature routinely, with periodic electrical and mechanical checks.

Motors, especially on floating aerators, can become coated with aerosols of the mixed liquor that dry out and form a surface coating. If the coating becomes too thick, it will act as an insulator on the motor's cooling fins and will lead to motor burnout if not removed routinely. Routine hosing of the motors should be a preventive maintenance activity and included in the preventive maintenance program.

Icing is another concern in some areas of the country. To prevent straining supports or overloading motors, ice should not be allowed to accumulate.

References

Eckenfelder, W.W., Jr., and O'Connor, D.J. (1961) *Biological Waste Treatment.* Pergammon Press, New York.

Houck, D.H., and Boone, A.G. (1981) *Survey and Evaluation of Fine Bubble Dome Diffuser Aeration Equipment.* EPA-600/2-81-222, U.S. EPA, Office Res. Dev., Cincinnati, Ohio.

Kennedy, T.J., and Boe, O.K. (1985) Efficient Aeration Operating Practices. Paper presented at 58th Annu. Conf. Water Pollut. Control Fed., Kansas City, Kans.

U.S. Environmental Protection Agency (1989) Fine Pore Aeration Systems. EPA/625/1-89/023.

Water Pollution Control Federation (1988). *Aeration.* Manual of Practice No. FD-13, Alexandria, Va.; Manuals and Reports on Engineering Practice No. 68, Am. Soc. Civ. Eng., New York.

Chapter 9
Solids Processes

INTRODUCTION

Wastewater sludge is the residuals left from the treatment of wastewater. Sludge is usually categorized by the process from which it is removed (for example, primary sludge, waste activated sludge, or digested sludge). Sludges removed by similar processes often have similar consistency, inorganic content, and dewaterability, but sludges removed from dissimilar processes generally have differing characteristics. From an energy perspective, the two most important characteristics of sludge are its water content and the enthalpy or heat content of its dry matter.

Water is difficult to remove from sludge and, beyond separation by simple settling (gravity thickening), generally requires chemicals or energy for further removal. The cost of removal of water increases exponentially with the water content of the sludge product desired (that is, the cost of drying is greater than the cost of dewatering, which is greater than the cost of thickening). Organic matter present in the sludge represents unburned fuel and, therefore, has inherent energy or heat value. The wastewater industry often refers to volatile suspended solids (VSS) as representing organic matter and, indirectly, heat content. In most heat calculation examples found in the literature, it would appear that the ratio of heat content to VSS was constant at 23 000 kJ per kilogram VSS (10 000 Btu/lb). In reality, the heat-content–to–VSS ratio will vary with the type of sludge and with the degree of decomposition already accomplished; that is, digested sludges have lower ratios of heat content to VSS than nondigested biological sludges, and biological sludges have lower ratios than raw sludges.

RECYCLE STREAMS. Minimization of contaminants in any sludge recycle stream is an important goal. Recycle streams not only return solids and biochemical oxygen demand back to wastewater treatment processes, but also can return unwanted nitrogen, phosphorus, and organic acids. Heavy recycle streams can reduce the treatment capacity of a wastewater treatment plant (WWTP), cause process upsets, and result in needless energy waste. Understanding the effect of recycle streams on treatment processes and knowing how to control them are essential to proper process control. Numerous books and technical papers have addressed these important issues.

Gravity thickeners should be operated to provide optimal concentration of solids without allowing needless aging of the sludge or excessive loss or washout of solids in the recycle stream. Dewatering units and chemical belt and drum thickeners should be operated to provide for a high solids capture rate while achieving maximum solids concentration and dewatering. Capture rates of 90% or more are achievable and should be targeted even at the expense of slightly less thickened solids.

PROCESS REMOVALS. In general, it is best to optimize the primary treatment process to save money and energy downstream. Care must be taken, however, to avoid too great a primary sludge age caused by either using too many primary clarifiers for the given flow or too deep a primary clarifier sludge blanket. This is especially important if cosettling with waste biological sludge is practiced. If organic sludge containing large populations of microorganisms is held for long periods, the first stage of anaerobic decomposition will begin to occur—acid fermentation with resultant drop in pH, reduction in alkalinity, and gas bubbling through release of carbon dioxide. Such conditions are typically detrimental to the performance of downstream biological processes.

Use of chemicals to improve removals in the primary clarifier can be of value. The tradeoff is in savings from less electrical energy used in downstream processes, increased gas production in anaerobic digesters, and, possibly, lower dewatering chemical cost and drier cake solids versus the cost of the chemical and any increase in costs associated with an increase of sludge as a result of the chemical itself.

In general, secondary treatment is achieved through biological oxidation of the raw or pretreated wastewater employing suspended-growth reactors (for example, activated sludge) or fixed film reactors (for example, trickling filter or rotating biological contactor). Suspended-growth reactors require that oxygen be delivered to the liquid process stream through compressors, blowers, or mixers; fixed film reactors require that oxygen be delivered to the wet process stream by pumping the liquid over fixed media, forming a thin film of liquid having a large surface area that is exposed to the air. Consideration must be given to the objectives of both the treatment process and the downstream solids processes when applying energy to the biological treatment process. Unless aerobic digestion is employed or desired, it is typically best to apply the minimum amount of oxygen in the secondary treatment process that will achieve the process objectives. This will minimize the use of electricity in the secondary treatment process while preserving the maximum amount of internal energy in the organic matter content of the sludge for downstream solids processes.

ANAEROBIC DIGESTION PROCESSES

In the anaerobic digestion process, organic matter is broken down to methane, carbon dioxide, ammonia, and water. This is a natural and relatively energy-efficient process. When properly applied, anaerobic treatment can result in a net production of energy, as opposed to alternative stabilization methods, which are essentially all net consumers of energy. Despite this, anaerobic processes are often not employed in applications where they are favorable, and many facilities that employ anaerobic treatment do not use the energy advantages of these processes to their full potential.

Anaerobic decomposition will occur at all temperatures between freezing and almost boiling, but optimum production of methane only occurs in the 32 to 35°C (90 to 95°F) mesophilic range and in the 54 to 57°C (130 to 135°F) thermophilic range. Methane-rich digester gas can be collected and burned to produce energy and/or heat. Because the conventional digestion process must be operated at approximately 35°C (95°F), the process itself requires a small amount of heat input as well as other energy input for mixing or recirculation. With the exception of cold winter periods in northern climates, properly designed and operated digesters should be net energy producers. Digester gas can be burned in engines to drive equipment or generate electricity. There is typically sufficient waste heat, which can be recovered from the engines to provide the necessary heat back to the digesters to maintain mesophilic

Table 9.1 Advantages and disadvantages of anaerobic stabilization processes

Advantages	Disadvantages
High degree of stabilization; inactivates pathogens	Slow growth rate of methanogens
Lessens amount of sludge for final disposal	Principal organisms (methanogens) have extreme sensitivity
Low nutrient requirements	Requires long solids retention times
Low energy requirements	May require auxiliary heating
Methane-rich gas is usable product	Capital intensive
Stabilized sludge is usable product	Maintenance intensive
	Generates a poor-quality sidestream

temperatures. Table 9.1 compares some of the advantages and disadvantages of anaerobic treatment.

TEMPERATURE. Anaerobic processes are typically operated in the mesophilic range, 32 to 38°C (90 to 100°F), although higher temperatures in the thermophilic range are sometimes used, and lower temperatures are possible at long solids retention times (SRTs). In general, higher temperatures result in more rapid decomposition of organics, thus decreasing the reactor volume required to achieve a specific degree of stabilization. However, there is seldom any benefit from a greater amount of VSS reduction because the optimum reductions of VSS within the two temperature ranges are typically comparable.

GAS COMPOSITION. Typically, digester gas is composed of up to 66% methane and 33% carbon dioxide by volume. However, the methane–to–carbon-dioxide ratio varies depending on the waste composition and on the alkalinity and pH of the digester contents. The ratio can be significantly different for various industrial wastes. It is important to be able to predict the distribution of the gases that may be produced in the design of systems for gas recovery, cleaning, and use.

ENERGY CONSUMPTION IN CONVENTIONAL DIGESTERS. The principal energy demands for conventional digesters are fuel for heating, pumped recirculation for heating, and mixing energy.

Digester Heating Requirements. Conventional digesters are typically operated in the mesophilic (32 to 38°C [90 to 100°F]) temperature range, although thermophilic (50 to 60°C [122 to 140°F]) treatment is sometimes used. Lower temperatures are seldom used because of the long SRTs required for stable operation. Accordingly, most conventional facilities are provided with supplemental sludge heaters. Generally, either natural gas or digester gas (or both) is used as supplemental fuel for sludge heaters, but fuel oil is also used. The amount of energy consumed is a function of the incoming waste sludge temperature, digester operating temperature, ambient temperature, and reactor construction and insulation.

The quantity of heat required for heating the influent liquid waste stream can be estimated as follows:

$$Q_s = W_s C_s (T - T_i) \tag{9.1}$$

Where

Q_s	=	heat required to heat influent waste to the digestion temperature, Btu/hr;
W_s	=	mass flow of wet influent sludge (mostly water), lb/hr;
C_s	=	specific heat of influent sludge, Btu/lb/°F (typically the same as water, 1.0);
T	=	digestion temperature, °F; and
T_i	=	influent waste temperature, °F (after heat exchange with effluent, if any).

Thus, assuming a typical winter influent temperature for municipal sludge of 10°C (50°F) and an operating temperature of 35°C (95°F), Q_s can be estimated as follows:

$$\begin{aligned} Q_s\ (\text{Btu/hr}) &= 8.34\ \text{lb/gal} \times q \times 1.0 \times (95 - 50) \\ Q_s &= 375\ \text{Btu/gal} \times q \end{aligned} \tag{9.2}$$

Where

q = sludge flow, gph.

Note that this calculated amount does not include efficiency losses from transfer of heat from the fuel source (for example, boiler and heat exchangers) or any benefit from heat exchange with the incoming sludge.

The heat required to compensate for heat losses through the reactor walls can be computed with Equation 9.3.

$$Q_r = UA(T - T_a) \tag{9.3}$$

Where

Q_r	=	heat loss to surroundings, Btu/hr;
U	=	composite coefficient of heat transfer, Btu/hr/sq ft/°F;
A	=	area normal (perpendicular) to heat flow, sq ft;
T	=	digestion temperature, °F; and
T_a	=	ambient temperature, °F.

The composite heat transfer coefficient per unit area, U, can be computed from the transfer coefficient of individual components in series with the direction of heat flow, as follows:

$$U = (1/U_1 + 1/U_2 + \ldots + 1/U_n)^{-1} \quad (9.4)$$

Where

U_1, U_2, and U_n = heat transfer coefficients for individual wall components between the sludge within the digester and the air or earth on the outside.

Table 9.2 lists heat transfer coefficients of typical anaerobic digester construction materials. The total heat loss through the digester walls is computed by summing the heat losses through areas of similar construction materials and temperature gradients (for example, roof or walls exposed to atmosphere, walls exposed to soil, and others).

Table 9.2 Heat transfer coefficients for various anaerobic digestion tank materials

Material	Heat-transfer coefficient, U, Btu/hr/sq ft/°F[a]
Fixed steel cover (0.25-in.[b] plate)	0.91
Fixed concrete cover (9-in. thick)	0.58
Floating cover (wood composition roof)	0.33
Concrete wall (12-in. thick) exposed to air	0.86
Concrete wall (12-in. thick), 1-in. air space and 4-in. brick	0.27
Concrete wall or floor (12-in. thick) exposed to wet earth (10-ft[c] thick)	0.11
Concrete wall or floor (12-in. thick) exposed to dry earth (10-ft thick)	0.06

[a] Btu/hr/sq ft × 3.155 = $J/m^2 \cdot s$; (°F − 32) × 0.555 6 = °C.
[b] in. × (2.540×10^{-2}) = m.
[c] ft × 0.3048 = m.

A typical heat loss value for the northern U.S. is 2.6 Btu/hr per cubic foot of reactor volume. This value is obviously lower in warmer climates. Typical heat loss values for middle and southern states are 1.3 and 1.0 Btu/hr/cu ft, respectively. With this background, typical heating requirements for single-stage sludge digesters in the northern U.S. are presented for various hydraulic loadings in Table 9.3, recognizing that some heat loss can be reduced by additional insulating.

Table 9.3 Supplemental heating required for single-stage completely mixed digesters in the northern U.S.

Hydraulic retention time, days	Heating of liquid from 50°F to 95°F,[a] Btu/gal[b]	Heat loss from digester, Btu/gal	Total digester heat loss, Btu/gal	Total heat required for heat losses,[c] Btu/gal
10	375	83	458	573
15	375	125	500	625
30	375	250	625	781
50	375	417	792	990

[a] (°F – 32) 0.555 6 = °C.
[b] Btu/gal × 278.7 = kJ/m^3.
[c] Fuel for heating corrected for energy efficiency of boiler at 80%.

The degree of stabilization is a function of sludge characteristics, digestion temperature, and SRT. In a completely mixed digester (assuming relatively uniform mixing), SRT is equivalent to the hydraulic retention time. Therefore, by concentrating the feed before digestion, both a substantial energy savings from sludge heating and a higher degree of stabilization from a longer holding time can be realized. In a well-mixed reactor, feed solids concentrations as high as 6% can be handled effectively within the reactor.

Energy Requirements for Sludge Heaters and Recirculation Pumping. Digester heating is generally provided by an external, boiler-heat exchanger, fueled either by digester gas or natural gas. Other, less popular methods are sometimes used (for example, direct steam injection and hot water pipes inside a digester). The efficiency of a sludge heater typically ranges from 70 to 85%. Higher efficiencies are associated with newer equipment that is well maintained. Heat exchanger scaling decreases efficiency significantly. In addition, electrical energy is required for pumped recirculation and miscellaneous water and oil pumps. Approximately 6 to 10 hp (4 500 to 7 500 W) is required per million Btu/hr heater capacity, corresponding to an electrical energy demand of approximately 6 kW/mil. Btu heat transferred.

Mixing Energy. Digester mixing is accomplished by pumping, or recirculating, the reactor contents. This is typically done using external pumps or compressors to recirculate gas or liquid, or using internal impeller mixers. Mixing is not used in standard-rate digesters or secondary digesters.

With respect to digester mixing, there is great disparity in mixer design and disagreement in perceived effect. In most installations, mixing is relatively ineffective: less than 50% of the total volume is effectively used. Early digester mixer designs were used for scum layer control; thus, they were not intended for blending the entire reactor contents. Digesters of the 1980s evolved from this technology and do not provide for effective mixing. However, based on mixing experience with heavy slurry suspensions in chemical process industries, it appears that an energy input of approximately 13 W/m^3 (67 hp/mil. gal) or greater of reactor content is necessary to provide effective mixing and prevent stratification. Table 9.4 summarizes digester electrical energy requirements for recirculation and mixing at this power level.

Table 9.4 Total energy requirements for completely mixed, conventional digesters

Hydraulic retention time, days	Electric energy for pumped recirculation, kWh/gal[a]	Electric energy for mixing, kWh/gal	Total electric energy, kWh/gal	Total electric energy,[b] Btu/gal
10	0.003	0.012	0.015	158
15	0.004	0.018	0.022	231
30	0.005	0.036	0.041	431
50	0.006	0.060	0.066	693

[a] kWh/gal × 951.1 = MJ/m^3.
[b] Equivalent heat required to produce indicated electrical energy at 33% conversion efficiency or 10 500 Btu/kWh.

Deposits of rags and grit in relatively flat-bottom type digesters result in reduced effective tank volume. Frequent cleaning is required to keep many digesters working effectively. This problem has led to the development and recent successes of the egg-shaped digesters. This shape allows for more effective recirculation and has proven to enhance the time–VSS-reduction relationship. While being more efficient from a mixing and VSS reduction perspective, egg-shaped digesters are more expensive to build and may have greater heat losses from their larger exposed surface areas.

ENERGY RECOVERY. Digester gas contains 40 to 75% methane, with 60% being common. Because methane has a higher heating value of 37 000 kJ/m^3 (1 000 Btu/cu ft), the higher heating value of digester gas is commonly taken to be 22 000 kJ/m^3 (600 Btu/cu ft). Approximately 0.75 to 1.25 m^3 of digester gas is produced per kilogram of VSS (12 to 20 cu ft/lb) destroyed in anaerobic digestion, with 17% being a common value. The latter is equivalent to a heating value of 23 000 kJ/kg (10 000 Btu/lb) VSS destroyed.

Digester gas is commonly used to provide the heat necessary to maintain digestion temperatures near 35°C (95°F). In areas where there is sufficient excess digester gas, installation of digester gas engines may be warranted. Engines can be used to drive pumps, blowers, or generators. Engine cooling water can be used as the source of digester heat when used in conjunction with appropriate heat exchangers.

As mentioned in Chapter 3, conversion of fuel to electricity is relatively inefficient. This is especially true when using a diluted fuel such as digester gas that is contaminated with carbon dioxide and hydrogen sulfide and is saturated with water vapor. Approximately 28% of the heat content of digester gas used in an engine generator goes to producing work, and at 90% generator conversion, only 25% goes to actual production of electricity.

For 1 ton (0.9 Mg) per day of VSS destroyed, the approximate value of electricity that can be produced in an engine-generator can be estimated as follows:

$$\begin{aligned} 2\,000 \text{ lb VSS/day} \times 10\,000 \text{ Btu/lb VSS} \times 25\% &= 5.0 \text{ million Btu/day} \\ \frac{5.0 \text{ million Btu/day}}{3\,415 \text{ Btu/kWh} \times 24 \text{ hr/day}} &= 61 \text{ kW} \end{aligned} \quad (9.5)$$

Therefore, for each 10 dry tons (9 Mg) per day of sludge with 80% VSS that undergoes 40% VSS reduction in an anaerobic digester, 195 kW (261 hp) of electricity can be generated.

$$\frac{10 \text{ ton/day} \times 2\,000 \text{ lb/ton} \times 80\% \text{ VSS} \times 40\% \text{ VSS red} \times 10\,000 \text{ Btu/lb VSS} \times 25\% \text{ efficiency}}{3\,415 \text{ Btu/kWh} \times 24 \text{ hr/day}} = 195 \text{ kW} \quad (9.5)$$

A more thorough discussion of digester gas energy recovery can be found in the U.S. Environmental Protection Agency's (U.S. EPA's) *Process Design Manual, Sludge Treatment and Disposal* (U.S. EPA, 1979).

AEROBIC DIGESTION

Aerobic digestion and anaerobic digestion are processes used to stabilize the organic matter in sludge. Unlike anaerobic digestion, which reduces sludge volatile matter through biochemical reduction processes, aerobic digestion reduces sludge volatile matter through biochemical oxidation processes using an aerobic environment. Aerobic digestion is energy intensive and produces a sludge that is often difficult and costly to dewater. It also has the disadvantage that it will undergo further anaerobic decomposition, with the production of odor, once oxygen is no longer available. A comparison of advantages and disadvantages of aerobic digestion is shown in Table 9.5.

Table 9.5 Advantages and disadvantages of aerobic stabilization processes

Advantages	Disadvantages
Moderate degree of stabilization; inactivates pathogens	Often produces heavy surface foam with asscciated odors
Lessens amount of sludge for final disposal	Sludge not fully stable; will further decompose anaerobically, producing odors
Fairly easy to operate	Requires long solids retention times
Less capital intensive and less maintenance intensive than anaerobic digestion	High energy requirements
Sidestream generated is not as poor quality as from anaerobic digestion	Sludge is difficult and costly to dewater

Wastewater treatment plants that use aerobic sludge digestion rarely use primary clarification because there is no economic benefit. Regardless of whether the primary solids are oxidized in the secondary treatment system or in the aerobic digester, approximately the same amount of oxygen and energy will still be required for oxidation. Primary clarification, then, becomes a needless extra step. Without primary clarification, however, the volume of an activated-sludge aeration tank will have to be increased to accommodate the increased loading. Because aerobic digesters are typically equipped with less efficient diffusers than the secondary process, it is often more desirable to complete as much oxidation of the primary solids as possible in the aeration system rather than in the aerobic digester.

Aerobic digestion of waste secondary sludge is a continuation of the endogenous respiration process that begins during secondary treatment. It requires the addition of oxygen through mechanical aerators or diffusers. Typically, the process requires 1 500 to 2 000 g of oxygen per kilogram of VSS destroyed (1.5 to 2.0 lb/lb). The higher value includes oxidation of the ammonia nitrogen released in the respiration process as well as the oxidation of the VSS. As proteinaceous material is broken down, nitrogen is released as ammonia and oxidized to nitrate in the aerobic digester. Oxygen transfer efficiencies will depend on the type of aeration device chosen and other factors as discussed in Chapter 8.

AUTOTHERMAL AEROBIC DIGESTION. Autothermal aerobic digestion (ATAD) is a modification of the conventional aerobic digestion process in which high concentrations of feed solids are employed in an insulated reactor. Aerobic digestion, like any other combustion process, is an exergonic process that releases energy in the form of heat. Heat is also generated in conventional digestion, but it is quickly dissipated in the higher volume of water associated with the sludge as a minor, almost imperceptible, temperature rise.

In ATAD the reactor volume is smaller, the feed solids concentration higher, the retention time lower, and the reactor enclosed and insulated. The result is a rising temperature that accelerates the digestion reaction. In colder climates, reactors may have to be provided with supplemental heat to maintain the desired temperature. Hydraulic retention times are generally on the order of 8 days but may be as high as 15 days.

While ATAD and thermophilic anaerobic digestion accelerate the digestion process, neither high temperature process has been found to result in any substantial increase in the amount of VSS reduced, or result in a substantial decrease in the overall amount of sludge produced, compared with lower temperature conventional processes. The main advantages of the high-temperature digestion process are that smaller reactors can be used and shorter retention times are required. Another advantage may be greater oxygen transfer efficiencies than with conventional aerobic digestion because of the high oxygen demand and greater rate of oxygen utilization.

It was reported by Kelly and Warren (1995) that ATAD reactors are not truly aerobic and produce offgases containing dimethyl sulfide and dimethyl disulfide (the same type of odorous byproducts given off by compost systems) as well as hydrogen sulfide. This may require that offgases from ATAD processes be scrubbed to remove odors, adding energy requirements and costs to the system input.

INCINERATION

Incineration is a process in which dewatered sludge is dried and burned to reduce the residual matter to inert ash having a greatly reduced volume. The process energy requirements vary significantly, depending on the type of sludge being processed and its moisture and organic content as it is fed into the incinerator, and on the air pollution requirements and operating practices. In most cases supplemental fuel is required to sustain combustion, but in all cases large quantities of heat are involved with great potential for energy recovery.

The most common types of incinerator are the multiple-hearth furnace (MHF) and the fluidized bed furnace (FBF). Generalized schematics for each of these incinerator types are shown in Figures 9.1 and 9.2. It is estimated that 231 MHFs and 37 FBFs are currently in operation for domestic sludge applications, and most of these were constructed before 1973 (Kelly and Warren, 1995). From an energy perspective, both types of incinerators exhibit

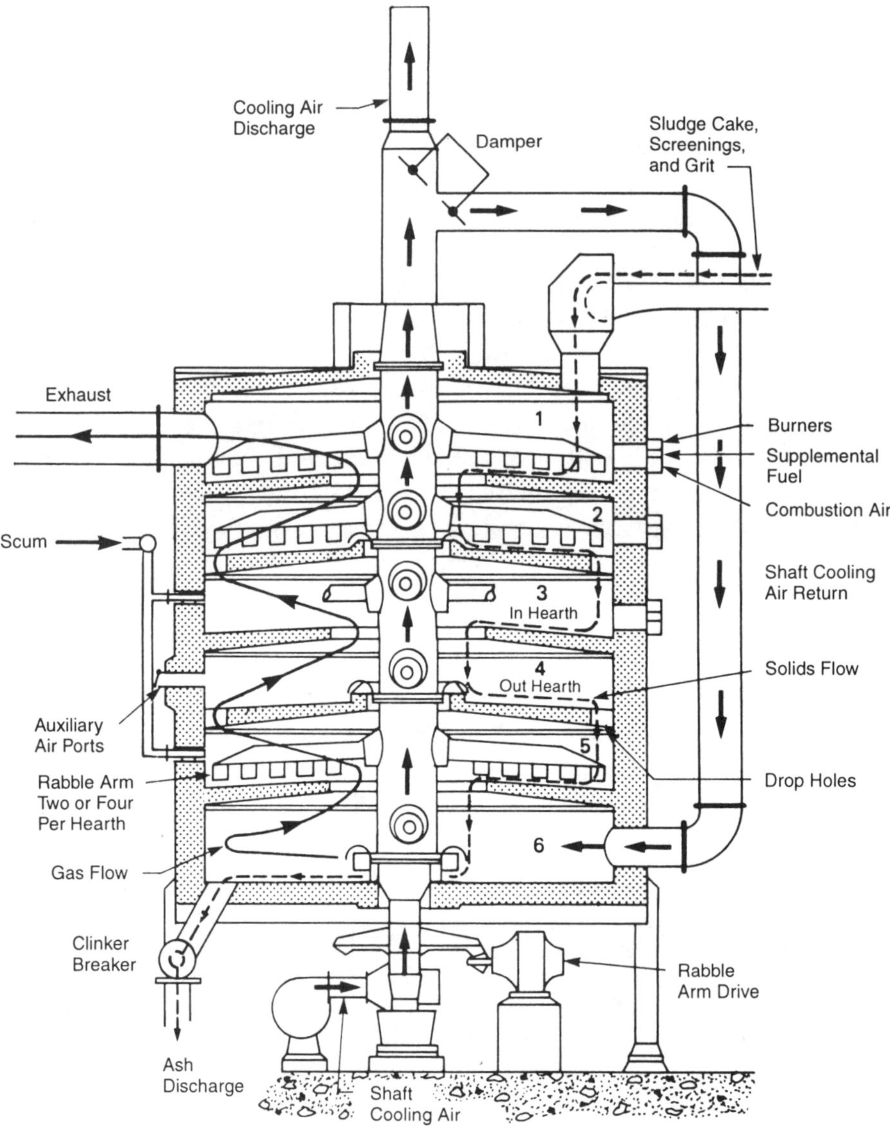

Figure 9.1 Cross section of a multiple-hearth furnace

many similarities: they start with the same type of wet sludge and end with the same type of ash, involve combustion of the sludge volatile matter with oxygen from air, involve the same type of thermodynamic principles, and have similar modes of heat loss. Operation of the MHF is somewhat more complex than that of the FBF in that sludge drying, combustion, and ash cooling take place at various levels in the incinerator as the material passes from top to bottom, while drying and combustion occur in the same single chamber of the FBF. Because drying occurs in the upper hearth of the MHF, the exhaust gas

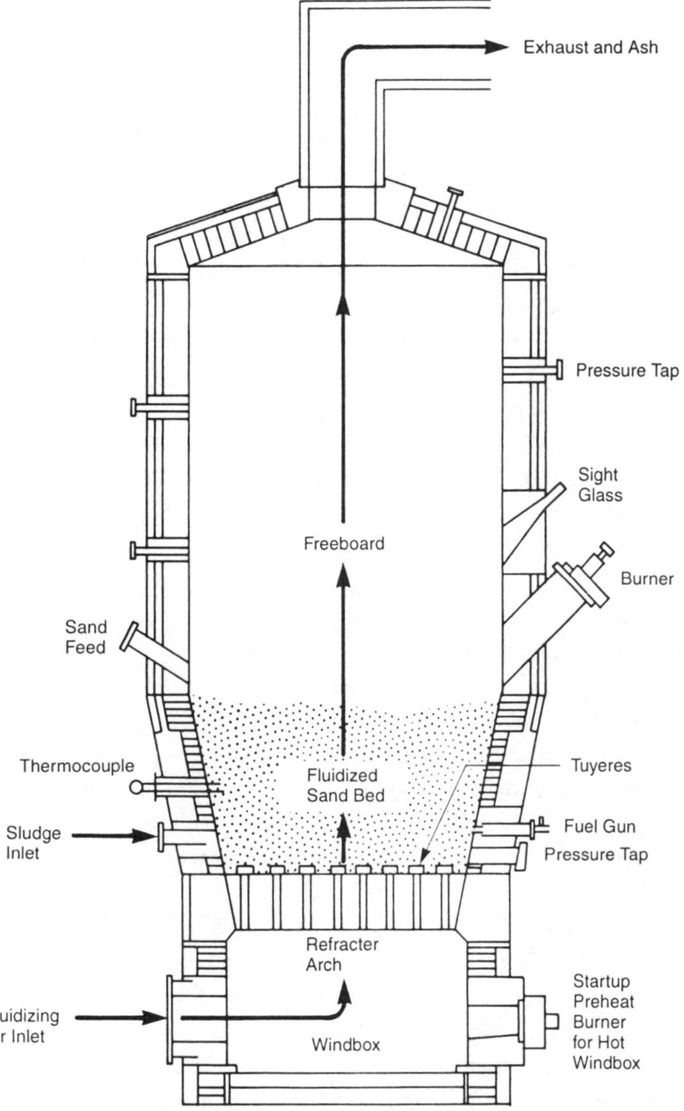

Figure 9.2 Cross section of a fluidized bed furnace

from the incinerator (without an afterburner) should have a temperature in the range of 500 to 650°C (900 to 1 200°F), which is significantly cooler than the temperature of the exhaust from the FBF, which is closer to combustion temperatures at 750 to 850°C (1 400 to 1 600°F). Consequently, to sustain this higher temperature, more heat is sometimes required for combustion in the FBF. However, the FBF will have fewer pollutants in the exhaust, and the potential for energy recovery from the exhaust gas can be greater.

FEASIBILITY OF INCINERATION. In many areas, the scarcity of available land has made siting for a landfill large enough to provide for 20 years of sludge disposal almost impossible. Incineration reduces the volume and weight of wet sludge cake by more than 90%, thus reducing the area required for residuals disposal. In recent years, advances have been made in the ability to recover heat from the incineration process and reuse the energy throughout the WWTP. This has increased the cost effectiveness of the incineration process. Recent improvements in thickening and dewatering technology have made it possible to obtain a drier cake from primary waste activated sludge mixtures. The drier cake reduces the auxiliary fuel requirements of the incineration process, making incineration less costly. Improvements in emissions-control technology have made it possible to meet existing emissions standards consistently (Walsh *et al.*, 1990).

PROCESS STABILITY. Stability and control are critical to the efficient operation of sludge combustion processes. One cause of process instability is an unsteady feed rate to the furnace, which results from variations in the performance of dewatering facilities and is exacerbated by the discharge of sludge cake from the dewatering system directly to the furnace feed conveyor, with no intermediate surge storage, such as occurs at MHF installations (Lewis *et al.*, 1989).

Because of possible blowback from the pressurized combustion chamber, fluidized bed installations have always had some degree of surge storage or *system capacitance,* such as a hopper above the sludge feed screw or pump. Providing a similar, though small, level of sludge feed surge storage (30 to 45 minutes) can improve the overall operation of an MHF installation (Lewis *et al.*, 1989).

The sludge inventory of an MHF is large, typically approaching 1 hour, whereas that of an FBF can be measured in seconds, and each hearth in an MHF has its own sludge inventory and is affected differently by process upsets. A greater or smaller sludge inventory within a furnace will not necessarily lead to improved process stability. However, changing the inventory of sludge over a relatively short period of time does affect process stability (Lewis *et al.*, 1989).

Multiple-Hearth Furnace. The MHF is the most common furnace in use for incinerating sludge. It involves feeding dewatered sludge at the top of the incinerator and allowing it to pass through the three stages of drying, combustion of organic matter, and cooling of the ash as it passes from the top hearth to the bottom hearth. The MHF is susceptible to upset and requires vigilant operational control. When either feed rate or moisture content changes abruptly, the steady-state conditions throughout the incinerator can quickly be thrown into imbalance.

Because of the irregular route of air flow through the incinerator, short-circuiting and poor air mixing can occur throughout the unit, requiring excesses of air for sludge combustion. This inefficient use of air results in a large heat loss for MHFs.

In February 1991, U.S. EPA published rules governing standards for the disposal of wastewater sludge by land application, land disposal, and incineration (40 CFR Part 503). These regulations require that exhaust gas oxygen analyzers be used with all MHF (and FBF) systems, allowing the integration of oxygen control into furnace operation. The 503 regulations also require that at least two independent thermocouples be provided in the combustion zone of MHFs. Because the combustion zone is dynamic, this will affect several hearths in the furnace. Even using two thermocouples may fail to provide reliable and representative temperature measurements unless the thermocouples are properly placed in the furnace and are of the appropriate insertion lengths (Sieger and Maroney, 1977).

Fluidized Bed Furnace. For FBFs, start-up fuel requirements are low, and little fuel is required for start-up following overnight shutdowns. In an FBF, the sand bed acts as a large heat reservoir, minimizing the amount of fuel required to reheat the system following shutdown, which makes FBFs good options for intermittent operation. Exhaust temperatures of FBFs exceed 750°C (1 400°F), so an afterburner, which requires supplemental fuel, to comply with air pollution regulations is often not required. Problems with FBFs include the feed system and temperature control with high-energy feeds such as dewatered scum (Lewis *et al.*, 1989).

Violent mixing in the fluidized bed ensures the rapid and uniform distribution of fuel and air and, consequently, good heat transfer and combustion. The bed itself provides substantial heat capacity, which helps to reduce short-term temperature fluctuations that may result from varying feed rates and feed heating values. Sludge particles remain in the sand bed until they are reduced to mineral ash. The violent motion of the sand in the bed grinds the ash material until it is so fine that it is readily stripped from the bed by the upflowing gases (Sieger and Maroney, 1977).

Even though FBF systems are typically more stable than MHF systems, process stability and control are linked to providing a steady rate of feed. In addition, the nature of the bed zone, because of the large heat sink capacity of the bed sand material, makes the development of an appropriate operating and control philosophy more complex. Bed temperatures tend to move deceptively slowly, often leading the operators to develop a false sense of security relative to normal operating routines and further complicating the configuration of control systems.

A key consideration in developing a reliable control philosophy for FBF systems centers on providing representative measurements of bed temperature. The 503 regulations require only a single thermocouple to be located in the bed zone; however, this is inadequate in most systems. While FBF systems provide excellent vertical mixing because of the turbulence of the sand bed, lateral mixing is relatively poor in most cases, making bed feed distribution an important design parameter. Even with good distribution, multiple thermocouples should be used at strategic elevations and locations around the perimeter of the bed (Lewis *et al.*, 1989).

HEAT REQUIREMENTS. Regardless of the type of incinerator used, heat balances are used to estimate energy use and requirements. The major heat inputs and losses that must be considered for incinerators are given below.

Heat gains (inputs) result from

- Combustion of sludge volatile matter,
- Combustion of auxiliary fuel, and
- Preheated combustion air (recirculated cooling air or heat exchange from exhaust).

Heat losses (outputs) include losses resulting from

- Water in sludge vaporized to exhaust,
- Exhaust of hot combustion gases,
- Exhaust of heated excess air and nitrogen associated with combustion air,
- Radiant heat from incinerator shell,
- Heated ash removed from the furnace,
- Vented shaft cooling air (MHF), and
- Gas cooling spray water vaporized to exhaust (FBF).

The major factors affecting auxiliary fuel usage are

1. Water in feed sludge cake,
2. Incinerator exhaust temperature,
3. Excess air for sludge combustion, and
4. Shaft cooling air (MHF only).

Energy losses associated with water are high and, therefore, significant regarding incinerator fuel usage and energy recovery. Although water is formed as a product of combustion and contained as humidity in combustion air, most water in the incinerator heat balance is from free water contained in the sludge.

Heat Losses Associated with Water. Because heat losses caused by water are so important in energy use and recovery, one should understand why these losses can be so large and energy recovery so poor. Even in the best dewatering applications, resulting sludge cake contains more water than solids. Rarely in wastewater applications will sludge feed cake in an incinerator have less than 60% water, and typically it will contain more than 75% water.

As the sludge is heated from ambient temperatures to exhaust temperatures, the heat capacity of its contained water undergoes three significant changes as the water is converted from liquid to gas (heat capacity represents the amount of heat required either to raise the temperature of a unit weight of the material or to transform a phase change). Water in the sludge remains as a liquid until its temperature is increased to the boiling point, at or near 100°C (212°F). At this point at standard atmospheric pressure, a large amount of heat is required to evaporate the water while the temperature of the liquid remains near 100°C (212°F). Once gasified, the temperature of the vaporized water can again increase. To determine the total heat required for heating water from ambient to exhaust temperatures, one must use steam tables or perform three separate heat calculations described as follows:

- Water as liquid—heating of the water (moisture) in the raw sludge from ambient to the boiling point of water. Approximately 1 Btu (1 kJ) is required to raise the temperature of 1 lb (0.5 kg) of the water in the raw sludge 1°F ([°F – 32]0.555 6 = °C) (this 1 Btu/lb/°F is the heat capacity of water).
- Water boiling—at the boiling point, water is converted from a liquid to a gas (steam) without a temperature change but requires a large input of heat energy. This phase change requires that 2 257 kJ per kilogram of water (970.3 Btu/lb) be supplied to convert the water from liquid to gas. (This 2 257 kJ/kg [970.3 Btu/lb] required for the phase change of water from liquid to gas is called the latent heat of vaporization.)
- Water as gas—once the water is converted to steam, more heat is required to raise the temperature of each pound of steam 1°F to the exhaust temperature. The heat capacity of steam (water as a gas) increases as its temperature rises, but is on the order of 0.47 Btu/lb/°F, less than half the heat capacity of water as a liquid.

Because the steam created is at atmospheric pressure, it cannot do useful work like the pressurized steam in a steam engine. Actually, because the exhaust temperature of the flue gas is rarely reduced to less than 200°C (400°F) with heat recovery devices, most of the heat energy used for water evaporation is wasted in the exhaust gas in the form of uncondensed steam. While it is technically possible to recover this heat, it is impractical for several reasons:

- Large heat exchanger surface areas would be required for condensation of the water vapor;
- To extract most of the heat, the temperature of any heat transfer medium would be below 100°C (212°F) and would be of marginal value; and

- The flue gas contains many impurities volatilized by heating of the sludge and auxiliary fuel. Strong acids, such as hydrochloric and sulfuric, and other impurities begin to condense before the water. This would require expensive acid-resistant condensing equipment if water were to be condensed to reclaim its heat.

Consequently, if a heat exchanger is used on the exhaust gas, only a small portion of the heat required to heat the water in the sludge from ambient to the furnace exhaust temperature would be recoverable. All of the heat used to heat the water from its initial temperature in the sludge to the heat exchanger exhaust temperature, including the heat of evaporation as a result of the phase change, will be lost to the atmosphere.

An example would be as follows (see Appendix for conversions). Assume initial feed sludge temperature of 60°F and an incinerator exhaust temperature of 1 000°F. The heat required to raise the temperature of 1 lb of water from 60 to 212°F would be

$$(212-60)°\text{F} \times 1\ \text{Btu/lb/°F} \times 1\ \text{lb} = 152\ \text{Btu}$$

The heat required for the phase change to evaporate 1 lb of water at 212°F and at 1 atm of pressure is 970 Btu. The heat required to heat 1 lb of steam from 212°F to the incinerator exhaust temperature of 1 000°F would be

$$(1\,000-212)°\text{F} \times 0.47\ \text{Btu/lb/°F} \times 1\ \text{lb} = 370\ \text{Btu}$$

The total is 1 492 Btu. Consequently, 1 492 Btu is required to heat 1 lb of water from 60°F to 1 000°F. Of this total, 1 122 Btu, or 75% of the total, is required to convert the water in the sludge to steam at 212°F. This leaves less than 25% of the total energy input (that portion used for heating the steam from 212°F to 1 000°F) available for energy recovery devices. The most important energy conservation measure, therefore, is to minimize the amount of water entering the incinerator because more than 1 100 Btu will be lost for each pound of water in the feed sludge without even considering other losses.

MULTIPLE-HEARTH FURNACE. When a MHF is placed on "hot standby," usually as the result of depletion of sludge inventory or as a result of manpower scheduling, it must be "buttoned up." This means closing all air ports, shutting down the induced draft fan, and closing off the shaft cooling air recycle, but not the cooling air. The furnace is maintained at a preset temperature of 400 to 500°C (800 to 1 000°F) through the use of auxiliary fuel. The only heat losses on hot standby are through radiation and convection, shaft cooling air losses, hot gases produced by auxiliary fuel, and air leaks to the idling furnace. All fuel used during hot standby is an energy waste, allowable only to protect the furnace refractory components. If the furnace were repeatedly cooled down to ambient and warmed back up to operating temperature, even at a slow rate of temperature change, there would be considerable abrasion of the firebrick as the brick expanded and contracted. This is of special concern for hearths because they have no internal supports other than their own self-supporting arch configurations. If too much abrasion or deterioration of the firebrick in the hearths occurs, the arch will fail and the hearth will collapse, typically causing additional failure of the shaft and rabble arms. Most furnace experts advise keeping the MHF "up to temperature" unless the planned outage is for needed maintenance or for a period of more than a week to avoid problems with the refractory materials in the furnace. Because hot standby is unavoidable, it is best to control air leakage to the furnace during idle periods and to monitor the temperature profile of the furnace. Ideal hot standby conditions would provide a temperature profile that is approximately 200 to 300°F lower than normal hearth temperatures, but individual hearth temperature control on standby is seldom achievable. Reduction of shaft cooling air volume may be of marginal benefit because a reduction in cooling air flow will be accompanied by a rise in exhaust cooling air temperature, which would produce a comparable heat loss.

Operations. If WWTP operators monitor the percent oxygen in the combustion equipment exhaust gases, they will have a better understanding of what is occurring in the equipment. If temperatures fall, an operator can observe the percent oxygen to determine whether the cause is an air leak or excess moisture in the sludge. Falling temperatures with an increase in excess oxygen indicate an air leak, while falling temperatures with no increase in excess oxygen indicate excess moisture in the sludge.

Good incineration requires complete combustion of sludge, efficient air pollution control, proper care of equipment, and safe operating methods. To achieve this, operators must understand how the incinerator operates, take special care in operating, and be able to respond to both routine changes and emergency situations.

Maintenance. A strong maintenance program increases the cost effectiveness and energy efficiency of the incineration process. It can reduce auxiliary fuel use by minimizing unscheduled maintenance shutdown and reduce costs for replacement materials by extending the useful life of furnace components such as the refractory brick work and other ancillary equipment. A strong program can provide the operators with the information necessary to operate the incinerators at high fuel efficiency by keeping the instrumentation and monitoring equipment operable and accurate.

AIR POLLUTION REGULATIONS. Air pollution regulations have become of paramount concern for incineration processes and currently dictate whether or not an incineration process will be operationally economical. The air permit rating of a furnace can become its de facto design rating, often limiting the sludge feed rate to a rate below the original furnace design. Air quality permits may require afterburners with high exhaust temperatures or multiple stages of air pollution control equipment, which add to capital and operating cost.

The 503 regulations, as enacted for sludge combustion, will primarily affect emissions of various metals and volatile organic compounds and, further, will require specific devices for monitoring key system operating parameters such as feed rate, exhaust oxygen, combustion process temperatures, and total hydrocarbons. In addition, other existing federal regulations affect emissions from wastewater sludge incinerators, and many state and local regulatory agencies either have in effect or are presently considering implementation of more stringent emissions regulations for these systems.

In reality, performance of the combustion and scrubbing systems cannot be optimized independently because the performance of both systems is unalterably linked to the stability of the overall process. Enhancing process stability must be considered a fundamental element in meeting stringent air pollution control regulations.

DRYERS

Sludge can be dried in mechanical dryers to

- Produce a dry pulverized material for use as a soil conditioner or fertilizer,
- Remove water before incineration, or
- Reduce the total volume and weight of the sludge before final disposal.

Dewatered sludge cake should be used as the dryer feed. Final cake moisture from the dryer depends on the intended use of the sludge following drying. *Scalping* is a term used when cake solids are only increased an additional 5 to 10%. This makes sense where waste heat, for example, from an incinerator can be used to provide evaporation of some of the cake moisture so that the feed sludge burns autogenously in the incinerator. For landfill disposal, intermediate dewatering to 50 to 65% cake solids may be appropriate. Without pelletizing, higher cake solids can result in severe dusting and a light, fluffy product, which among other problems becomes difficult to handle. Some of the commercially available dryers are specially designed to produce a granular or pelletized product that is low in dust.

Sludge enters a glue or sticky phase when it gets above 30 to 40%, depending on the type of sludge. To prevent the sludge from sticking to the dryer parts, some of the finished material must be recycled back and mixed with the feed sludge to make the feed material more friable.

PURPOSE. Dryers have been used to produce a dry material containing organic matter and nutrients that can be used as low-grade fertilizer and soil conditioner, or supplemented with nutrients to produce a commercial-quality fertilizer. Generally, the sludge used in the dryer is biologically unstabilized and, once wetted, becomes biologically active, produces heat, and often produces odors. Nevertheless, a large amount of this dried wastewater sludge originating from large U.S. cities has been used for agricultural purposes for many years.

Dryers do not necessarily have to remove all the water in the sludge and are often used to remove only a portion of the water. This can be done by partial drying of the entire sludge stream or by combining wet sludge with dried sludge to attain the desired product. Both the costs of landfill disposal and incineration can be reduced by partial drying of sludge.

Where landfill space is expensive, as it has been in the northeastern U.S., it can often be found to be economically viable to dry sludge before disposing of it in a landfill. Because it takes approximately 75 L (20 gal) of fuel oil to evaporate 0.9 Mg (1 ton) of water, drying can be an attractive option even in the absence of waste heat.

Sludge that is dried to a high degree is often voluminous, dusty, and difficult to handle. For landfill disposal, it is generally more feasible to dry the sludge to solids of 80% or less for ease in handling.

Because drying is part of the incineration process, it is, in most cases, more efficient and cost effective to design an incinerator to do both tasks rather than using a combination dryer/incinerator. The exceptions to this are

- Existing installations that are currently limited in incineration capacity because of high water content of sludge. In this instance, it may be found that adding a dryer to reduce a portion of the sludge water can maximize production of capital already in place. Production of the incinerator can be increased by predrying all or a portion of the sludge.
- Incinerator installations that have afterburners with heat recovery and no useful purpose for the recovered heat. Heat that would otherwise go to waste could be used for drying of the sludge to make the overall process more efficient and more economical.
- Anaerobic digestion operations in which surplus gas can be used to provide heat for drying.

In general, it will be found that separate drying of sludge is inefficient and not cost effective when used in conjunction with incineration. However, there are specific applications for which this is not true. These applications are typically the result of unusual local or economic conditions or limitations of existing expensive incineration equipment.

Heat balance calculations for dryers are similar to those for incineration. Because there is no heat generated in the process, most of the heat loss is attributable to the amount of water in the sludge to be dried. Recovery of heat from the dryer flue gas is often impractical and seldom attempted.

Dryers have significant heat and electrical requirements. In addition to a heat source, dryers require pumps or blowers for circulating the heat transfer medium as well as drives, mixers, conveyors, and other auxiliary devices.

HEAT TREATMENT

During the 1960s and 1970s, many heat treatment units were installed in the U.S. These were generally of two types: pressurized with air and pressurized without air. Many are still in use today, but new installations are uncommon. Both types worked in principal like household kitchen pressure cookers and both types broke down complex and refractory organic matter under heat and pressure to simpler materials. The units pressurized with air were also of two types: low pressure and high pressure. The high-pressure units also operated at higher temperatures and achieved a reasonably high degree of oxidation of organic matter, whereas only minor amounts of organic matter are oxidized in the low-pressure units. The more common low-pressure units are considered to be a means of conditioning the sludge for dewatering rather than of destroying the sludge solids. Nevertheless, even the low-pressure oxidation units could be relied on to destroy a small portion of the sludge, often on the order of approximately 10%. Heat-treated sludge from any of the various processes can generally be dewatered to 35% or more through simple dewatering processes such as vacuum filtration, belt filter pressing, or centrifugation.

The largest waste of energy in operating any of these units is from trying to process too low a concentration of feed sludge because most of the heat is lost in the exiting processed liquid as elevated temperature. Most of the heat used in this process is wasted, with only a small fraction being recovered for preheating of the sludge. Consequently, nearly half as much heat energy would be required to process the same quantity of sludge at a 4% feed concentration as compared to a 2% concentration. This, however, is only true to a limited extent because electric energy use per cubic metre (gallon) does increase with increasing feed solids concentration as a result of increased backpressure. As solids concentration exceeds 5%, the energy required for pumping and backpressure increases dramatically.

One of the most significant energy concerns for heat treatment processes involves the retreatment of the organic material solubilized through heat and pressure that is returned to the main treatment process. The decant or filtrate contains a high concentration of soluble biochemical oxygen demand, chemical oxygen demand, and ammonia-nitrogen, which is often returned intermittently at a high rate for short durations.

REFERENCES

Kelly, H., and Warren, R. (1995) What's in a Name?—Flexibility. *Water Environ. Technol.,* **7,** 7, 46.

Lewis, F.M., *et al.* (1989) Design, Upgrading and Operation of Multiple Hearth and Fluidized Bed Incinerators to Meet the EPA 503 and Other Proposed New Regulations Presented at Sludge Composting, Incineration and Land Application: Burning Issues and Down-to-Earth Answers. Virginia Water Pollut. Control Assoc., Richmond, Va.

Sieger, R.B., and Maroney, P.M. (1977) Incineration-Pyrolysis of Wastewater Treatment Plant Sludges.

U.S. Environmental Protection Agency (1979) *Process Design Manual, Sludge Treatment and Disposal.* Munic. Environ. Res. Lab., Cincinnati, Ohio.

Walsh, M.J., *et al.* (1990) Energy-Efficient Municipal Sludge Incineration. *Water Environ. Technol.,* **10,** 36.

40 CFR Part 503. Code of Federal Regulations, Washington, D.C.

Chapter 10
Energy Management

ENERGY MANAGEMENT PLAN

Electricity (energy in general) is either the largest or second-largest expense in a wastewater treatment plant (WWTP) budget alongside personnel costs. Many WWTP managers and supervisors have little or no control over personnel costs but typically have absolute control over energy use. Therefore, cost control through energy management should be a high priority for all WWTP managers.

Typically, the first approach taken when it is decided to reduce energy use is to make a casual WWTP survey to determine if and where there are excesses and abuses. Lights are turned off, thermostats are adjusted, minor changes are made in operation, and possibly a more thorough look at WWTP operations is initiated. Often, process changes are even implemented that result in reduced blower or aerator usage, less dissolved air flotation system use, less pumping, and other reductions. However, without ongoing management awareness, all successes achieved in these efforts disappear in time because the operation typically drifts back to what had been considered normal, safe, or easy.

Many suggestions for reducing energy consumption at WWTPs have been published, and many of these suggestions can be put into practice. However, for continuing success, there must be an energy management program.

Energy management is not as complicated as it might seem. Once it has been determined that energy management is necessary, a program can be set up with four simple steps: gather data, analyze data, create a plan, and implement the plan.

The first two steps pertain to the energy audit. They define what exists, what current operating practices are, and where and how energy is being used. This information provides the information needed for further decisions. The third step sets the direction for the program and determines whether additional expenditures are required for new or replacement meters or equipment or for professional assistance. The last step is implementing the plan. The following describes in more detail the steps for an electrical energy management program.

GATHER DATA. The first step in preparing an electrical energy management plan is to conduct a WWTP audit in which all WWTP electrical equipment is inventoried. Data to be collected would include name of the equipment, nameplate information such as motor power and revolutions per minute, a description of whether the motor has a constant load, whether it is operated full time or part time, whether it is of constant speed or variable speed, and whether it has a variable load. Other data would relate to heating, ventilation, and air conditioning equipment, outdoor lighting, chlorine evaporators, laboratory equipment, and other factors. A good portion of this step may already be done in many WWTPs through the maintenance management system.

Gathering data consists of taking inventory of WWTP electrical equipment and listing nameplate data on all electric motors of at least 745 W (1 hp) and other equipment of at least 1 kW. During the WWTP audit, it is typically best to start at each motor control center (MCC) and itemize each piece of equipment in order as listed on the MCC. Also, itemize all electric meters on MCCs and local control panels (such as kilowatt meters, ampere meters, and equipment hour meters).

The appropriate design data for a driven element, such as a pump's or blower's capacity, as well as motor efficiency can be obtained from equipment submittals, which should be on file at the WWTP. Determine whether a pump or blower is associated with a flow meter, pressure gauges, an ammeter, or other equipment. Gather information for operating and billing records. Operating records along with run-time meters can be used to determine frequency of use for WWTP equipment and create a historical comparison of energy use. Usage records from the local electric or gas utility should also be collected. The local utility will provide a summary of electric use and demand for each billing period of the last 12 months at no charge on official request.

A qualified electrician should check the power draw of each major piece of equipment. Appropriate readings would include ammeter, voltmeter, wattmeter, and power factor readings (as equipment is available) for large motors under full load, as well as partial load for variable output equipment. Flow meter and pressure gauge readings as available should also be made, along with the various electrical readings. An instrument mechanic should verify the accuracy of all panel meters.

Data gathering is an endless task. Operator log sheets should be checked to determine whether operators are recording appropriate energy management information such as electric meter totalizer readings, run-time readings, and ammeter readings on at least a once-per-week basis. If not, forms should be developed for listing the various meter readings not being recorded. In most cases it may be appropriate to create a new log sheet for entering energy management data. It is best to organize entries in the order in which the new data will be collected, for example, in order of entry on MCCs if hour meter readings will be taken from there. Data can be collected daily, weekly, or monthly. Weekly may be preferred because it smooths out daily aberrations, and potential problems can be detected early. In addition to electrical readings and hour meter readings, data to be collected include influent flows, returned activated sludge flows, blower supply, and analytical data including mixed liquor volatile suspended solids and biochemical oxygen demand (BOD) data for appropriate wastewater streams. While BOD data may not be as timely as other information, it is still worthwhile to compare past aeration system electric use with actual WWTP loadings.

ANALYZE DATA. Once the raw data are collected, they must be properly analyzed. One of the first steps to electric energy management is creating an electrical budget using historical information. With the data collected, estimates of usage for the past measurement period are chronicled and compared with the electrical use recorded on the WWTP wattmeter. Using the budget analogy, the wattmeter can be thought of as measuring income (incoming) and the individual equipment usage as expenditures (outgoing). Once this budget has been adequately tested to verify where all the electrical energy is being used, one can determine where and how electrical energy use can be controlled or reduced. The budget approach should be continuous, comparing weekly usage with the forecast usage.

There are various approaches to estimating actual electrical usage in the absence of wattmeters on all pieces of equipment. For constant-speed equipment, it is often easiest to use run-time meters and power draw for each piece of equipment. For pumps and variable-load equipment, run-time meters are seldom of any use, and other estimating methods must be used.

Pumps. Pumps consume energy in relation to the amount of fluid pumped. Obtain the equipment manufacturer's literature and determine the pump's power ratio (kW/mgd); convert horsepower to kilowatts by multiplying horsepower by 0.746. Compare manufacturer's data with field-obtained data as available. If field data indicate that a pump is not producing full flow according to the manufacturer's literature, determine whether the cause is worn wear rings. Otherwise, use the ratio determined by the most reliable field information. In the absence of any other data, a quick approach is to take the operations and maintenance manual or design criteria data for the pump and divide the maximum pump capacity by its horsepower.

In all of the following examples, it will be assumed that the motor efficiency and the loading on the motor are approximately the same so that the motor input is equivalent to the motor nameplate (output) rating.

Example: Design criteria for a main wastewater pump list the pump criteria as 3 700 gpm at 50 hp (see Appendix for conversions to metric). Convert the design criteria data to power per unit flow, as follows:

$$50 \text{ hp} \times 0.746 \text{ kW/hp} = 37.3 \text{ kW}$$

$$\frac{3\,700 \text{ gpm} \times 1\,440 \text{ min/day}}{1\,000\,000 \text{ gal/mil. gal}} = 5.33 \text{ mgd}$$

$$\frac{37.3 \text{ kW}}{5.33 \text{ mgd}} = 7.00 \text{ kW/mgd}$$

AERATION EQUIPMENT. Aeration equipment typically represents the greatest energy-consuming items in a WWTP. The power draw of aeration equipment is difficult to estimate. Equipment power draw is a variable that should be related to the BOD loading to the aeration system and to the food-to-microorganism (F:M) ratio under which the system is being operated (see Chapter 8 for further discussion). However, to be able to directly relate power draw to BOD loading, it is necessary to know the oxygen-transfer efficiency of the system, which will depend on the type of aeration device used and its operating condition. Actual measured power consumption of the aeration equipment needs to be measured and compared with the oxygen requirements as calculated in Equation 10.1. Actual oxygen-transfer efficiencies are in the range of 0.8 to 1.8 lb O_2/hp-hr (lb × 0.453 6 = kg; hp-hr × 2.685 = MJ). The lower value is for inefficient coarse-bubble-type diffusers, whereas the higher value is for the most efficient fine-bubble (fine-pore) diffusers. These numbers are considerably lower than would be found in the literature or from manufacturers.

Example: A coarse-bubble diffuser system with a known field oxygen-transfer efficiency of 1.0 lb O_2/hp-hr is being used to treat 1 000 lb/d (5 000 mg/s) of primary effluent BOD at a F:M ratio of 0.3 (see Appendix for metric conversions).

$$R = 0.75 + [0.056/(\mathrm{F{:}M})] \qquad (10.1)$$

$$R = 0.75 + [0.056/0.3] = 0.94 \text{ lb } O_2/\text{lb BOD}$$

$$O_2 \text{ required} = 0.94 \text{ lb } O_2/\text{lb BOD} \times 1\,000 \text{ lb/d BOD}$$
$$= 940 \text{ lb } O_2/\text{day}$$

$$\text{horsepower required} = 940 \text{ lb } O_2/\text{d}/1.0 \text{ lb } O_2/\text{hp-hr}/24 \text{ hr/d}$$
$$\text{horsepower required} = 39.2 \text{ hp (or} = 29.2 \text{ kW)}$$

SOLIDS SYSTEM EQUIPMENT. Equipment used in the thickening, dewatering, and stabilization of sludge is often operated intermittently in relation to the amount of sludge to be processed. Often, solids system equipment consists of many small motors for drives, conveyors, pumps, and so on. It is typically best to record run time for the system and use a summary of the system's connected power.

DEVELOPING AN ENERGY MANAGEMENT PLAN. The ideal energy management plan would include a kilowatt meter indicating instantaneous kilowatt draw and a kilowatt meter that totals kilowatt use over time for each major piece of electrical equipment or for each group of identical equipment. All such meters should be tied into the WWTP's main control panel or other central location along with a readout of the main WWTP kilowatt meter. This would give the operator in charge a complete view of electrical usage in the WWTP. Alarm points or direct computer control programs could be added as refinements.

A less sophisticated plan can also be effective that makes use of existing WWTP meters that are commonly found at most WWTPs. This plan uses the WWTP kilowatt meter, flow meters, equipment hour meters, and panel ammeters. This plan can be executed using a personal computer and a spread-sheet program.

Neither hour meters nor a computer is essential. Run times of intermittent operations can be ascertained with fair reliability from operator log sheets as long as operators are properly instructed in logging on–off times of all significant electrical equipment that they manually place in service. In many cases, even with hour meters, estimates must be used. For example, a continuous-duty air compressor that loads and unloads with demand may indicate 24-hour-per-day use but may never be near its full load rating for more than a small fraction of the day. Lastly, although a computer simplifies calculations, a manually developed plan should take less than 1 hour per day in calculation time.

IMPLEMENT PLAN. Energy management plans can take on many different forms, but the most important feature of any plan is the implementation stage. Audits and studies of energy usage should pinpoint areas of both high energy usage and inefficient energy usage. Plans may include specifics of capital projects to recommend and fund, but they should also provide for continuous monitoring of usage and for a means to see when inefficiencies arise.

Peak Electric Demand Reduction. As discussed in Chapter 3, an electric bill for a typical WWTP consists of four major components: customer charge, demand charge, energy charge, and fuel surcharge.

This chapter concentrates on the demand charge portion of the bill as it relates to peak electric demand reduction. The other three components mentioned above are explained in Chapter 3.

Demand is defined as the rate of using electrical energy. For example, ten 100-W lamps burning at the same time require 1 000 W (1 kW) of electricity-generating capacity or demand. Burning for 1 hour, the lamps would consume 3.5 MJ (1 kWh) of electric energy.

Demand charge is the charge billed to a customer for the utility company to maintain the generation, transmission, substation, and distribution capacities needed to meet the maximum demand. Maximum demand is the simultaneous demand imposed on these facilities owned by the utility company from the different users during a certain given time interval. The demand charge is based on a rate set by the utility company, in their rate structure, that is multiplied by the demand reading as measured *and as defined in the rate structure.*

The demand reading is the highest electrical demand registered in a set time interval (typically 15, 30, or 60 minutes) during the billing period.

Demand reading is measured either in kilowatts or kilovolt amperes (kVA). In billing on a kilovolt-ampere basis, all power used, including nonworking power, is measured and billed, while billing on a kilowatt basis only includes the working power. Nonworking power is the power required to produce magnetic fields needed for operation of any inductive load, such as motors or transformers. A surcharge may be levied for a low power factor when using the kilowatt basis. The difference between the kilowatt basis and the kilovolt-amperes basis is related to power factor, as defined in Chapter 3. A low power factor indicates a large difference between kilovolt-amperes and kilowatts and wasted electrical power. The objective should be to maintain the highest power factor possible for high electrical efficiency. Smaller demand capacity customers are typically charged based on kilowatts. Larger demand capacity customers are typically charged based on kilovolt-amperes. This discussion will use kilowatts because that is the most common term for wastewater treatment systems. Actual demand results in the demand reading mentioned above.

The demand charge is designed so that the customer pays a fair share of the utility's fixed investment in production, transmission, and distribution equipment required to meet the customer's maximum requirements. Typical single-family residential meters do not record kilowatt demand; they record only kilowatt-hour (megajoule) consumption. Electric heat residential customers, however, may be required by the utility to measure demand. Again, the convention is that the higher the impact on the utility demand base, the more charges there are for demand-based rate application.

A utility company bases its demand charge on the highest amount of electricity consumed by the customer during the demand measurement interval: the more electricity used at any given time, the larger the possibility of an increase in a utility company's investment in generation, transmission, and distribution systems. One way of obtaining a reasonable return on these systems would be to average all generation, distribution, and transmission equipment costs among all customers. That approach would not be fair, however, because those who use the utility's equipment at a steady rate would be subsidizing those who do not. Although the peak demand periods are typically of a short duration, just like WWTP peak loadings, the reserve capacity must be available.

Demand is not *instantaneous* demand, as many people assume. Starting up a large motor does not result in peak demand for calculating demand charge. Demand is the average power required during an established demand interval. The most commonly used demand intervals are 15 minutes and 30 minutes, but 60 minutes and other time intervals are used by some companies. The typical demand register records average energy consumption for each 15- or 30-minute interval in a day. When the first interval ends, the equipment resets and starts on the second one. The utility reviews the demand records at the end of each billing period. In most cases, the maximum demand recorded is used to compute the demand charge. There also is a ratchet clause in some rate structures that states that charges are to be based on the highest demand achieved over a period of 6 or 12 months, even if the present month peak demand is lower than the previous peak demand. Utility companies having problems in meeting the system-demand requirements will generally devise the rate structures with a ratchet clause. These ratchet clauses are what make peak demand control cost effective.

In some cases, the ratchet clause applies only to a certain percentage of the previous peak demand. In some cases, the peak demand is measured according to on-peak hours and off-peak hours.

The methods of demand charge and rate calculation are unique to each power company and must be understood before a program aimed at peak demand reduction is initiated. Generally, power companies are willing to explain the billing methods and will probably offer suggestions for reducing the demand charges. On request, utilities can often supply a demand record by intervals. An analysis of this record will identify the periods of peak demand. An investigation of operations during peak periods generally reveals activities that can be deferred until nonpeak periods. For example, testing of backup pumps and blowers during a peak-load interval is expensive. Money could be saved by testing and using electric equipment with high-energy demand during off-peak times. Another method is to use an alternate power source or storage facility for stormwater during the peak demand period.

Figure 10.1 shows bar graphs of months versus peak demand for Hudson Tinkers WWTP-5 and Hudson Terex WWTP-6. Hudson Tinkers WWTP-5 has a peak demand occurrence in one month that is almost 100% more than previous peak demand; however, Hudson Terex WWTP-6 has a peak demand that is only 12% higher. A high peak occurred at Hudson Tinkers WWTP-5 in the month of September because of the second pump that came on, which usually works as a standby pump only. Stormwater conditions triggered the second pump to turn on. An on-site standby power system, which is required by U.S. EPA in some cases, or higher wet well level start for the second pump if the system can store all the wastewater without causing basements to flood could have prevented this peak from occurring. Significant cost savings can be realized by using one of the methods discussed in the following sections and by equalizing the demand over the entire year for the Hudson Tinkers WWTP-5 facility.

In predominantly peak-causing billing, in which frequent simultaneous starting and stopping of equipment occurs in lieu of continuous running of equipment, the demand charge portion of the bill could easily exceed 50% of the total charges. Demand charges generally make up 25% of the total electric bill for a typical WWTP.

While the power company rates mentioned above were used in developing the models for determining demand/energy consumption and cost, it is important to note that the rate structures are constantly under review and changed often. For example, the power company could examine the effects of changing

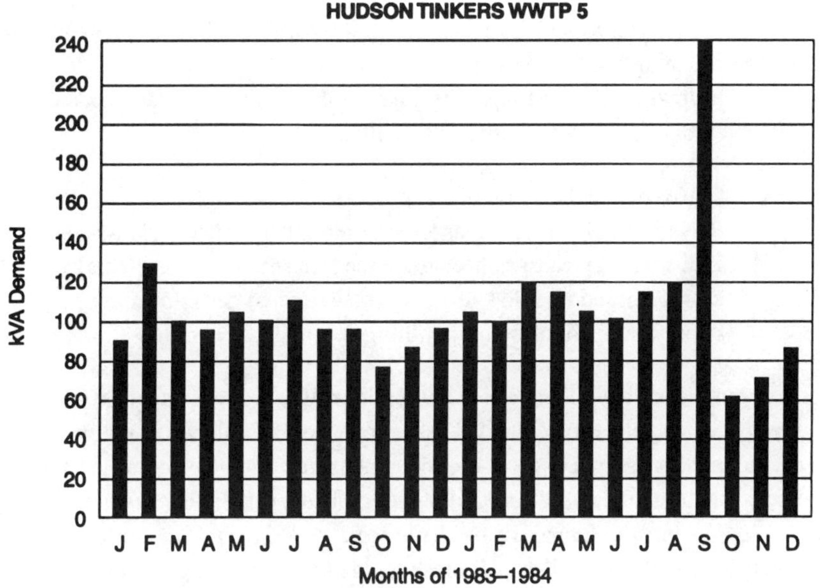

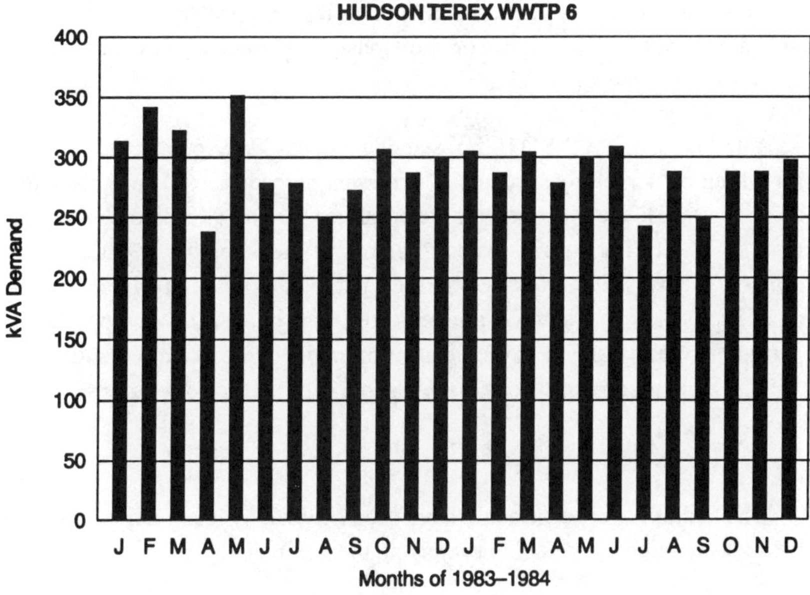

Figure 10.1 Months versus peak demand at a wastewater treatment facility

the 60% ratchet to 80% or 100% in a given rate schedule. For this reason, the effects of rate changes at different locations should be examined on a regular basis.

In conclusion, the three major methods of peak electric demand reduction being presented here should be implemented after thorough investigation of billing data, rate structure, effect on remaining portions of the WWTP, cost effectiveness, operational requirements, and feasibility of the implementation of a peak demand reduction system.

The following listed billing advantages could be realized if a peak electric demand reduction system were implemented:

- Lower demand portion of a given month's bill,
- Prevention of demand charges stemming from ratchet clauses, and
- Lower energy charges because some rate structures have energy charges based on kilowatt-hours (megajoules) per kilowatt or kilovolt-ampere of monthly billing demand.

A number of techniques and combinations of techniques can be used to lower the peak electric demand. These range from simple yet effective low-cost manual programs to complex and expensive computer-controlled systems. The basic idea used for load reduction is the same: *Operate only essential equipment simultaneously, with the least expensive power source.* Load identification and reduction methods include load surveying, load shedding, load staggering, off-peak usage, and on-site engine or power generation.

Any or all of these may be useful in establishing a peak electric demand reduction program that can, if conscientiously implemented, save substantial amounts in the energy budget.

EXAMPLE 1. A 7 500-W (10-hp) standby process water pump is exercised for 1 hour on a monthly schedule. The normal procedure is to turn this pump on in addition to the other process water pumps that are already on line during regular service. The wire-to-water efficiency of the pump is 74%; thus, it draws 10.1 kW. If it is exercised during a peak-load period, it will contribute all of its 10.1 kW to the maximum demand charge of \$3.72/kW, imposed for a period of 1 year. Calculated on the maximum, the pump exercise would cost \$451 in demand expense alone (see Appendix for metric conversions).

Demand cost = 10.1 kW × \$3.72/kW × 12 mo/yr = \$451/yr

This additional expense could have been avoided by using one of the methods mentioned, including exercising the pump at night.

EXAMPLE 2. For larger systems, demand effect will be greater. Hudson Tinkers WWTP-5, having a demand increase of 110 kVA, will result in much larger demand cost to the facility; that is,

Demand cost (August)	=	120 × \$11.35/kVA
	=	\$1 362

Demand cost (September) = 240 × $11.35/kVA
= $2 724

Demand cost as measured (October) = 60 × $11.35/kVA
= $681

Billable demand cost if 100%
ratchet clause (October) = $2 724

Additional demand cost because of
100% ratchet clause (October) = $2 724 – $681
= $2 043

Calculate for 60% ratchet (0.6 × $2 724) – $681
= $1 634 – $681

Additional demand cost because of
60% ratchet clause (October) = $953

The additional billing described above could occur for the next 5 or 11 months, depending on the rate structure. Assuming that for these succeeding months, the actual measured demand remains at 60 kVA:

Total annual additional demand cost
(6 months with 100% ratchet) = $2 043 × 6 = $12 258

Total annual additional demand cost
(12 months with 100% ratchet) = $2 043 × 12 = $24 516

Flow Equalization. Wastewater treatment plants are typically designed and built to treat a fluctuating flow stream, one that varies widely throughout the day and possibly the season in both quantity and strength. These variations in loading require that process units and equipment be large enough to meet the daily peak demand, seasonal peak demands such as rain events, and in some cases the projected demand imposed by future growth. Peak loading is generally of short duration; therefore, facilities operate at less than design capacity most of the time. By controlling the flow to the process at a more even rate, loadings are more consistent, and biological and energy demands are more stable. Through the dampening effects of equalization, only the treatment units and equipment needed to meet the average demand are required to operate. The result is an overall improvement in WWTP efficiencies, more consistent removal rates, reduced peak-demand charges, and possibly decreased power consumption. Additional benefits include the dissipation of shock loads that most WWTPs experience and extension of the operating capacity of the existing facility because the initial design capacity was oversized to allow for peak demands.

Equalization basins provide one method of equalizing the change in demand loading on the system. Equalization basins are sized according to the diurnal variations in loading for a specific facility. The volume is generally less than 35% of the WWTP capacity. They can be located on site or upstream of the WWTP and arranged as in-line or off-line tanks. Off-line basins or tanks are most widely used. They are constructed of a variety of materials; lined earthen basins are one common form. At underloaded WWTPs with design capacities intended to provide service to a growing community, some part of the original tankage can probably be used for flow equalization with only minor modifications. The opportunity also exists at upgraded or expanded facilities having abandoned tankage to convert unused tanks into equalization basins with minimal capital expenditures.

Aeration and/or mixing may be required to avoid septic conditions or settling. Generally, this entails a low power demand but must be considered when considering flow equalization.

Equalization typically affects only raw wastewater pumping, and the effect generally is not large; it depends on the amount of runoff and presence of infiltration into the sewers or manholes. Analysis may need to be made to rehabilitate sewers and manholes versus using one of the peak-shaving methods described herein.

Priority Load Shedding. Priority load shedding reduces peak demand by turning off or keeping off loads that are noncritical to the process or operation of the system during the peak demand periods. Priority load shedding can be achieved by manual means or by using automatic control systems.

For example, wasting of sludge, if it is not detrimental to the process, can be accomplished during the nonpeak or nighttime operation, when peak demand charges or ratchet clause effects are smaller. Also, aerators or aeration blowers can be used in cycle fashion in lieu of turning them all on and running them together.

Scheduling of various loads in order of priority with shedding of noncritical loads is called *priority load shedding,* or *load management.* Load management is the scheduling or control of electrically powered equipment to minimize peak electric demands. The two techniques addressed in this section for meeting this objective involve optimizing WWTP operations, scheduling, and establishing a set of actions to take if demand exceeds a predetermined level.

Once WWTP operations schedules have been optimized, a plan of action should be developed to further reduce electrical demands, contingent on the actual demands experienced at the WWTP. This particular aspect of load management is often automated to avoid operator errors, provide greater vigilance, and free the operator for other tasks. Automation is not essential, however. A wide range of automated load management (or energy management) systems are available. Some of the major system types and their functions are addressed here. A growing trend is toward the use of systems that not only automate load management functions, but control and schedule major WWTP operations as well. Automatic systems can be stand-alone systems, available in the market today from numerous manufacturers, or they can be incorporated in an overall computer control system for WWTP process control. Features will vary from manufacturer to manufacturer and model to model; however, the factors discussed in this chapter need to be included when purchasing or specifying the equipment.

Feedback signal for these systems can be measured by two different methods. Separate current transformers to measure current and potential transformers to measure the voltage can be installed at a main service location to calculate kilovolt-amperes using the following formulas:

For kVA-based rate structure,

$$\text{kVA} = \frac{\sqrt{3} \times \text{Voltage} \times \text{Current}}{1\,000} \tag{10.2}$$

For the kilowatt-based rate structure, the power factor must be measured using a power factor transducer. Use the following formula for kilowatts:

$$\text{kW} = \frac{\sqrt{3} \times \text{Voltage} \times \text{Current} \times \text{Power factor}}{1\,000} \tag{10.3}$$

The second approach, which is the most preferred, less expensive, and most accurate, is to ask the utility company servicing the location for the pulse input from their metering system. This pulse input, with appropriate multipliers, can be connected directly to the load management system. This input will represent actual kilovolt-amperes or kilowatts being metered by the power company.

In addition to understanding the basic functions of load management systems, it is important to know the basic steps in developing a load management plan. The following steps are recommended to help develop a load management plan:

1. Determine the electrical load imposed by various pieces of equipment and rank them by importance and ability to be cycled.
2. Establish electrical demand goals and demand levels to trigger corrective action.
3. Provide some means by which to compare actual on-line electrical demands with the goals and targets established under item 2 above.
4. Specify a sequence of actions to be taken if demand trigger thresholds are reached.
5. Determine the amount of time the load can be kept off.

Only when these steps have been accomplished can the decision and design of automated equipment be determined. It should be noted that load management systems are only applicable when the facility is demand-metered and has numerous discretionary loads (loads that can be turned off).

There are a number of calculations involved to determine the cost-effective application of automated load management systems. It is advantageous to use spreadsheet-type software packages to develop a customized spreadsheet. Different rates can be applied to this spreadsheet to arrive at data that will enable a decision to be made for a comprehensive load management system.

Priority load shedding or load management involves setting up a contingency table in which various pieces of equipment are ranked in order of expendability. The most expendable pieces of equipment can then be targeted for "turn off" as demand levels reach the appropriate trigger point.

Data for maximum shed time, minimum shed time, minimum restore time, and the priority number are all derived from design and operation experience of the particular WWTP. These values are not calculated but determined from the answers to the following questions:

1. Which process equipment is least important to the process?
2. Which process equipment is most important to the process?
3. What is the maximum and minimum amount of time process equipment can be operated without affecting WWTP operations adversely?
4. The WWTP cannot run without what equipment? This equipment should not be connected to the load management system.

Horsepower or kilowatt data, which are either design data or data from the nameplate of the equipment, are important in determining the total kilowatt amount being shed.

ON-SITE ENGINE OR POWER GENERATION UTILIZATION

If either flow equalization or peak electric demand reduction is not feasible, then on-site engine or power generation should be considered.

Most WWTPs require an emergency source of mechanical or electrical power, as mandated by U.S. EPA. This source is required for the treatment process operation during a power interruption from the power company supplying power under normal operating conditions. The emergency source of power could be an engine-driven pump, a second electrical power feed, or an on-site generating system. The power company typically charges an exorbitant amount to provide a backup power feed. Because of this, it is generally economical to provide an on-site power-generating system. In addition, the option of using this on-site generating system for peak shaving or peak electric demand reduction should be considered.

There are four different design/application methods of using on-site power-generating systems:

- Engine-driven pump,
- Power-generating system—traditional transfer scheme,
- Power-generating system—synchronized transfer scheme, and
- Power-generating system—parallel with utility.

While considering this option, engine/generator operation and maintenance costs should be considered against the savings that could be realized from peak electric demand reduction.

ENGINE-DRIVEN PUMP. This method is inexpensive to implement. If the engine-driven pump is required as mandated by U.S. EPA for backup purposes, then capital cost to provide the pump should not be included in the payback analysis to be performed for peak electric demand reduction. It is also assumed in this method that because most WWTPs are manned 24 hours/day, additional manpower to run the pump during peak electric demand periods is unnecessary.

Dual-driven systems can be used to increase the overall reliability. An electric motor and a right angle drive for the engine can both drive the pump, alternately. During high-demand periods, the engine can be turned on and used to drive the pump, thus saving peak electric demand costs.

After analyzing the need for peak electric demand reduction, the pump can be turned on either manually or automatically during the peak electric demand cycle. If peak-electric-demand–causing periods are predictable, then exercising of the pump can be synchronized with the peak demand occurrence. Standby pumps typically must be exercised under load once per week.

The subsequently described methods are used more often than the engine-driven pump, mainly because of the lack of flexibility of power usage and standby electrical power unavailability.

POWER-GENERATING SYSTEM—TRADITIONAL TRANSFER SCHEME. In this option, a power-generating system is used.

All of the above methods achieve the same goal: provide power for peak electric demand reduction. This goal is achieved at different costs and savings depending on the method used. Along with cost, the convenience of using a system that is adaptable to a particular application should be considered. This option provides the flexibility of selecting any load in the system rather than only one that has an engine connected to it, as previously discussed.

When considering this option, the power-generating system should be designed to provide necessary peak power for the duration that peak power is necessary. It may be necessary to derate the generating unit because of the longer continuous running time required compared to the standby application.

A traditional transfer scheme will create total power interruption. It is recommended to provide dead time between the power-transfer operation to ensure that back electromagnetic forces from the deenergized load have been completely decayed. In-phase monitors are available as an option if total power interruption is to be avoided.

This method is the safest and least complex of the three being described in this section.

POWER-GENERATING SYSTEM—SYNCHRONIZED TRANSFER SCHEME. In this method, during the generation of power, the power system grid and the generating system are synchronized and allowed to operate in parallel. Power is then fed into the WWTP electrical distribution system as required by the grid demand.

The advantage of this method is that the load does not have to be turned off during the peak shaving cycle, which would cause a process interruption. A synchronized transfer scheme will keep wastewater pumps running during the transfer cycle, providing closed transition transfer, and will not take the pumping system off line.

It may be necessary to use these types of systems when the process upset is not acceptable.

POWER-GENERATING SYSTEM—PARALLEL WITH UTILITY. This method is similar to the synchronized transfer scheme. The major advantage in this method is that if the city or village that owns and operates the WWTP also owns and operates the electric utility, then the electric-grid-wide peak electric demand reduction could be realized by operating the generating system.

On a hot summer day, when all air conditioning units are running and brownouts are occurring in the city electric utility system (because of heavy demand from the customers, and with demand at the WWTP not that high), the generating system could be turned on, parallel with the electric utility grid, and power exported to the grid to lower the systemwide demand.

Payback could be calculated by comparing the cost of the switchgear and the equipment necessary to provide the paralleling system against the benefits realized in lower demand and related ratchet clause effect.

SELF-GENERATION OPTIONS

Self-generation options vary widely but can be discussed generally in terms of three primary self-generation options available to WWTPs:

- Using emergency backup generators for peak shaving;
- Installing natural gas cogeneration units; and
- Using digester gas, sludge, and other byproducts of the wastewater treatment process as fuels.

USING EMERGENCY BACKUP GENERATORS FOR PEAK SHAVING. Nearly every WWTP has backup generation capability of some type for emergency situations. Depending on the structure of electric rate options provided by the local utility, peak shaving may be a viable method of achieving significant reductions in power costs with minimal incremental investment.

The concept of peak shaving is to offset peak energy and demand purchases by self-generating during peak (highest rate) periods. This may take the form of self-generating all power required during such periods or shaving the amount of peak purchases from the local utility by supplementing peak-period purchases with self-generated power.

Because emergency generators are already installed and operational, the only additional costs generally incurred would be in the following areas:

- Direct fuel and supply costs,
- Environmental permits authorizing the facility to operate at higher levels of output, and
- Staff time and training required to operate the equipment at an increased frequency.

While this is an attractive approach to reducing power costs, it may not always be feasible to implement. One reason is that the local utility may not be supportive of the customer's decision to self-generate during high-revenue periods and may assess unreasonably high demand rates to attempt to compensate for the expected loss in revenues. Another problem is that standby generators using diesel fuels may not be allowed by the local air pollution control district (APCD) to operate on a nonemergency basis.

Should this be the case, the WWTP should investigate one of the other two options described in this chapter.

INSTALLING COGENERATION UNITS. Another approach might be to install one or more cogeneration units to supply the facility's energy needs with power purchased from the local utility only during emergency or down periods. The economics of this option depend on a variety of factors:

- The size, capacity, and associated capital investment of generating units required to serve the facility's electric requirements;
- The availability of financing, and rates and terms;
- The cost and availability of fuel resources;
- Operation and maintenance costs (such as labor, equipment, maintenance, and allowance for capital repairs);
- The ability of the WWTP to use thermal energy produced by the cogeneration equipment, and the corresponding economic value for offsetting thermal energy purchases; and
- Opportunities, if any, for the WWTP to sell excess power produced to the local utility or another power consumer, and the expected revenues produced by such excess power sales.

To assess the viability of such an option, a comprehensive cost-benefit analysis should be performed, incorporating engineering design criteria and identifying optional configurations. This kind of study should be performed by a specialist in this field who is familiar with both the technical aspects of cogeneration and the economic analysis of cogeneration projects.

USING DIGESTER GAS, SLUDGE, AND OTHER BYPRODUCTS OF THE WASTEWATER TREATMENT PROCESS AS FUELS. A variation of the cogeneration theme would entail using digester gas, sludge, and other byproducts of the wastewater treatment process as fuels. This approach can significantly enhance the economic benefits realized by offsetting or eliminating the need for purchased fuels. Additional key economic variables are

- Capital investment;
- Financing terms;
- Cost and availability of fuel;
- Operation and maintenance costs;
- Value of thermal energy produced; and
- Potential revenue from sales of excess power.

Any savings realized from reduced waste disposal costs should also be factored into the analysis.

Obviously, the power-generation facility alternatives can be quite complex. The optimal configuration ultimately depends on operating and design characteristics specific to the WWTP, subject to engineering design, financing, and fuel resource constraints. Other complicating factors include constraints imposed by regulatory agencies, such as environmental permitting.

FEASIBILITY EVALUATION

An evaluation of options for saving energy costs entails an analysis of a variety of factors specific to the WWTP's circumstance. The primary factors to consider are listed below. It is important to note that no one factor should be considered independently of the others because they are all interrelated. The project configuration selected will ultimately be that which represents the optimal choice, given the WWTP's objectives for this project; the environmental, economic, and other constraints; and the ratio of benefits to costs.

RANKING OF PROJECT OBJECTIVES. Before embarking on any project, the WWTP must identify and prioritize the project objectives. These may include

- Serving the WWTP's minimum, average, peak load;
- Using waste products from the treatment process;
- Maximizing financial benefits (through offsets of retail power purchases and/or sales of excess power);
- Constructing a showcase project demonstrating the viability of new technology or the feasibility of reemploying waste products for productive purposes; or
- Any combination of the above.

SITING FACTORS. For a planned self-generation project, siting of the power-generation facility is generally made on the basis of several factors, including the following:

- Proximity to the fuel source(s),
- Location of the load(s) being served,
- Disruptions and efficiency of WWTP operations,
- Location of nearest commercial and/or residential communities,
- Distance from utility distribution lines (if excess power is to be sold), and/or
- Space available.

Other factors that may enter into siting considerations are the following:

- Environmental—if the WWTP site is near a residential area, concerns about noise, smells, and other types of pollution may affect the location of the generation plant.
- Economic—when more than one choice of site is available, the relative costs versus benefits of the respective options may be the determining factor. For example, all other factors being equal, if the WWTP owner has the option of siting the generation facility near the fuel source or

near the utility distribution line and these are at different locations, the most economic approach in terms of capital costs may be the determining factor.

DESIGN FACTORS. The design selection process involves two basic steps: determining appropriate system size and operating mode, and identifying prime mover type.

Again, a variety of factors interrelate with the design selection decision. The most important element at this stage of the analysis is to ascertain the WWTP's power requirements, both thermal and electric, and its demand and energy profile on a time-of-use (daily, weekly, monthly, and seasonal) basis.

The WWTP's power requirements then provide the basis for several sizing approaches:

- Sizing the generation facility to supply the WWTP's minimum load;
- Sizing the generation facility to supply the WWTP's average load;
- Sizing the generation facility to supply the WWTP's peak load; or
- Sizing the generation facility to optimize the use of nonconventional fuel resources (for example, sludge, digester gas, and/or landfill gas).

For each facility size to be evaluated, one or more prime mover types should be selected on the basis of fuel resources available, capital cost considerations, and significant performance characteristics (for example, conversion efficiency and emissions).

Each project configuration (consisting of facility sizes and prime mover types to be evaluated) should then be evaluated on the basis of its ability to meet technical, economic, operational, and environmental project objectives.

ECONOMIC FACTORS. The economic evaluation includes consideration of a variety of factors, leading to a decision regarding the optimal project size and configuration. Primary factors include

- Wastewater treatment plant current and projected power requirements (electric and thermal, demand, and energy);
- Cost of utility-provided power, current, and forecast;
- Value of excess power generated (if an outside market exists);
- Cost and availability of fuel resources (for example, natural gas, diesel, fuel oil, digester gas, landfill gas, or wastewater sludge);
- Site costs (procurement, rental, and lease);
- Capital costs (including preliminary design, engineering design, permitting, equipment, construction, construction supervision, and cost of capital for the various design options considered);
- Capital cost credits (for example, capital costs offset by constructing the project, such as the cost of air emissions control equipment for

digester gas, which need not be purchased when the digester gas is used as fuel);

- Cost of air emission offsets purchased and required (if any) to operate the facility;
- Operation and maintenance costs (whether performed by staff or outside contractor);
- Savings attainable by avoiding disposal and other costs (for example, disposal costs avoided by using sludge as fuel);
- Capital replacement cost budget (for scheduled and nonscheduled replacements);
- Tax benefits available (such as investment credits and alternative fuel credits);
- Cost of capital (whether interest paid on project debt or facility's cost of capital inclusive of equity, on a discounted time-value-of-money basis); and/or
- Proportion of equity contribution and project risks and benefits to flow to each project participant (including facility owner, developer, and financer).

All of the above factors should be quantified and input to a spreadsheet modeling the project's expected financial performance on a long-term basis (10 or more years, depending on factors such as the estimated project life, financing and contract terms, and quantity of fuel resources available). The project may be evaluated on the basis of

- The value of discounted net cash flows over some term;
- Whether or not it generates an acceptable rate of return for the facility owner and other project participants;
- The extent to which it is expected to achieve other objectives (for example, environmental and technological); or
- Any other set of criteria established by the project developers.

OPERATIONAL FACTORS. The operational requirements of any planned generation project must also be considered. Operational factors include

- The number and types of staff required to operate the generation facility,
- Staff scheduling and training,
- Frequency of equipment maintenance, and
- Fuel processing.

ENVIRONMENTAL FACTORS. A variety of environmental factors must also be considered when planning a power-generation project. These vary from one geographic location to another and are subject to local, state, and/or federal regulations. Consequently, the respective environmental regulatory agencies and other local authorities should be consulted for an update of the environmental constraints applicable to the project.

Although the specific requirements vary widely, several types of environmental constraints are fairly common to all generation projects in the U.S.:

- General—U.S. EPA establishes federal guidelines for protection of the environment, which are then delegated to the respective states for implementation. These regulations have resulted in a series of regulatory approvals governing construction of new facilities such as power-generation facilities. One of the most important of these is the environmental impact report (EIR), which must be prepared by a new facility. The EIR is a comprehensive report documenting the planned facility design and the types of effects it is expected to have on the environment. Environmental impacts include
  - Air quality (such as organic and inorganic emissions, visibility, and smell);
  - Noise;
  - Water quality (such as chemical content and temperature);
  - Population density (such as effect on traffic patterns, parking, and community services); and
  - Cultural, historical, scenic, and other characteristics of the proposed site.

  Virtually any expected effect of any kind on the environment (flora and fauna) and the adjacent community must be documented, reported, and assessed before a project can be approved. In addition to review by one or more regulatory bodies, the EIR is also subject to public review and comment.
- Air pollution—air emissions are subject to regulation by the U.S. EPA, state air resources boards, and regional APCDs. U.S. EPA establishes minimum air quality requirements for each state and air basin. The respective states then establish programs to attain and/or maintain these minimum federal requirements. States may also opt to impose more restrictive requirements to attain a higher air quality standard than that required under federal law. It then becomes the responsibility of the local APCDs to administer the state program for their specific air basins.

The following is a listing of some permits required by U.S. EPA and/or the APCDs for construction of a new facility, such as a power generation facility, under new source review (NSR) procedures:

- Authority to construct—the APCD will generally require the developer to submit preliminary design plans and equipment specifications, including estimated quantity and type of pollutants emitted given the planned fuel mix. The APCD will then evaluate the data provided by the developer in terms of whether or not the facility meets local limits on certain types of pollutants. The APCD may require the project developer to apply best available control technology (BACT) to mitigate the adverse effect of controlled pollutants on the environment.

- Prevention of significant deterioration (PSD)—projects exceeding certain size and/or emissions constraints are subject to additional scrutiny through the PSD review process. The purpose of the PSD is to ensure that new sources of pollutants do not have a significant negative effect on the air quality of the region. The PSD requires that the project developer model air emissions, taking into account a variety of site-specific factors (for example, wind currents, ambient air temperatures, and height and points of emissions) to demonstrate that the project will not have a deleterious effect on regional air quality. In addition, all major new sources are subject to BACT.
- Permit to operate—on completion of the facility, the APCD may require emissions testing to confirm that the facility will perform within emissions limits established in the authority to construct. A permit to operate will then be issued, authorizing the owner to operate the facility, subject to certain conditions (for example, fuel constraints, emissions constraints, and operating hours) for a specified period of time. The facility owner may also be required to perform periodic emissions testing to retain the right to operate.

A variety of environmental and other permits (for example, building) may be required, based on local regulatory objectives and constraints. Of these, air emissions constraints tend to be the most significant when determining the feasibility of a power-generation facility of this kind.

However, in the event that hazardous wastes are being considered as fuel resources (for example, use of landfill gas from a superfund site), the project may be subject to other permits and approvals (for example, health risk assessment that evaluates the risk to neighboring communities as a result of the emission of known carcinogens). The key environmental factors affecting feasibility of each project configuration must be evaluated in the context of its unique mix of resources, technology, performance, and other characteristics.

FINANCING APPROACHES

The developer can finance development, construction, and/or operations of an electric-generation plant. The following are types of financing available:

- Construction loans—short-term financing during construction that may be refinanced with another, usually lower interest, loan upon commencement of operations;

- Working capital loans—short-term financing to provide working capital during development, construction, and/or operations; and
- Fixed-term loans—loans of a fixed duration for some specified purpose, for example, as refinancing of a construction loan.

The specific financial structure used depends on the financing objectives and the amount of risk and equity to be assumed by the project developer. The primary types of financial structures are revenue bonds, conventional bank financing, lease financing, privatization, and joint ownership and/or development. The financial structure put in place will depend in large part on the amount of risk assumed, equity to be contributed, and benefits to be allocated to the respective project participants and financers.

This section focuses on the concept of project financing in the context of the primary financial structures listed above.

PROJECT FINANCING. Project financing is a method of financing capital-intensive construction projects, wherein the project and its assets, contracts, operating revenues, and cash flows are evaluated by a lender as a separate entity. The primary appeal of project financing is that the debt may be nonrecourse: that is, if the developer can demonstrate that project cash flows may be reasonably expected to support all expenditures including operations and maintenance and debt service, the developer may not be held responsible for meeting debt service obligations during periods when project cash flows fall short of expectations. Further, because the project financing is dependent on the financial viability of the project alone, the debt capacity of the developer is not affected.

Any one or combination of financial structures listed above can be used for project financing. The key element is that the project economic structure be able to stand on its own.

REVENUE BONDS. Most WWTPs are owned and operated by municipalities or special districts; thus, tax-exempt bonds are the primary means of financing construction of an electric-generation plant. If the WWTP is privately owned, the developer may be able to issue industrial development bonds or industrial revenue bonds. Industrial development bonds and industrial revenue bonds may be either taxable or tax exempt. Pollution control facilities such as WWTPs, including local power-generation facilities servicing such WWTPs, are tax exempt for industrial revenue bond purposes.

When revenue bonds are issued, the revenue stream from the generation plant is committed to repayment of the bonds. Special funds (for example, operations and maintenance, surplus, capital replacement, and capital reserve), financial management, and reporting requirements are established by the bond covenants to ensure protection of the bondholders' investments in the plant. The specific interest rate and terms available are dependent on the debtor's financial position and bond rating as well as financial market conditions.

CONVENTIONAL BANK FINANCING. If public debt is not a viable option, commercial bank loans may be explored. These may be in the form of secured or unsecured loans and may involve one or more lenders.

The specific terms available will vary widely with the credit standing of the developer, the financial strength of the project, and financial market conditions. In general, the following characteristics will apply:

- Rate(s) of interest, whether fixed or variable, will be specified for a stated term.
- The lender is entitled to repayment of the loan with full recourse (that is, the borrower is ultimately responsible for debt service, regardless of whether or not the project meets projected cash flows).
- The lender does not share in project risks or benefits.

To qualify for a loan, the developer will likely be required to contribute some percentage of equity, demonstrate the overall financial viability of the planned project, and pledge securities to protect the lender in the event of default on the loan.

LEASE FINANCING. Another option to explore would be lease financing. Lease financing is a means of financing large equipment purchases and projects in which ownership of the leased property is a key issue. The full range of leasing options is complex and not appropriate for discussion here. The primary types of leases can be summarized as follows.

Direct Financing Lease. Lessor provides all funds necessary to purchase the leased asset (also referred to as nonleveraged and capital leases), and lessee assumes all of the risks and benefits of ownership. Direct financing leases have one or more of the following characteristics:

- Ownership of the leased asset transfers to the lessee at the end of the lease term;
- Contains a bargain purchase option, allowing the lessee to purchase the asset at a price less than the fair market value at the end of the lease term;
- Term of lease equal to or exceeding 75% of the estimated useful life of the leased asset; and
- Present value of lease payments, including minimum payments during any noncancellable terms, equal to or exceeding 90% of the fair market value of the leased asset, less any investment or other tax credits retained by the lessor.

Leveraged Lease. This is similar to a direct lease but involves a minimum of three parties: lessee, lessor, and long-term lender. There are many variations on leveraged leases, but all leveraged leases have the following characteristics:

- The lessor assumes responsibility for repayment of the loan from the lender(s) and assumes ownership of the project. The loan is secured with a first lien on the equipment, assignment of the lease, and assignment of lease rental payments. The financing provided by the lender(s) is substantial to the transaction and nonrecourse to the lessor.
- The lessor's net investment declines during the early years of the lease term and increases during the later years of the lease term.
- Investment and other tax credits retained by the lessor are accounted for as one of the cash flow components of the lease.

Operating Lease. An operating lease is a means of providing financing that is not considered a purchase by the lessee (also known as *off balance sheet financing* because operating lease payments are considered rents; the associated lease obligation is not recorded on the lessor's balance sheet and does not reduce the lessor's borrowing capacity). Operating leases are defined as those that do not meet the criteria of capital or direct financing leases above; that is,

- Ownership of the leased asset *does not* transfer to the lessee at the end of the lease term.
- The lease *does not* contain a bargain purchase option allowing the lessee to purchase the asset at a price less than the fair market value at the end of the lease term.
- The term of the lease is less than 75% of the estimated useful life of the leased asset.
- The present value of the lease payments, including minimum payments during any noncancellable terms, is less than 90% of the fair market value of the leased asset, less any investment or other tax credits retained by the lessor.

Conditional Sale Lease. A conditional sale lease is a means of installment purchase financing sometimes available from large equipment manufacturers in which ownership is transferred to the lessee with a bargain purchase option at less than the asset's fair market value. All risks and benefits of ownership of the leased asset are assumed by the lessee. A conditional sale lease is classified as a capital lease for accounting purposes (that is, the lease obligation must be recorded on the lessee's balance sheet as a liability).

Certificates of Participation. Certificates of participation are a form of public financing that have the appearance of tax-exempt revenue bonds but are structured as leases. Basically, title to the property being financed is held by certificate holders and amortized over the term of the certificates. Payment on certificates is guaranteed by revenue generated by the financed facility. At the end of the lease term, ownership reverts to the municipality or public agency that issued the certificates.

Tax-Exempt Leases. Tax-exempt leases are direct leases financing construction of public facilities and other qualifying facilities constructed for the benefit of the public. Interest revenues earned on tax-exempt leases are not taxable. When tax-exempt debt is employed, any available tax credits (such as investment, energy, and nonconventional fuels) may have to be foregone. Therefore, benefits gained from tax-exempt debt must be evaluated against the cost of tax benefits foregone, if any.

Sale-Leaseback. In a sale-leaseback, the owner of the asset sells the property to a third party (lessor) who immediately leases the property back to the owner (lessee). The leaseback may be either a capital lease or an operating lease. If it is a capital lease, project financing results.

Lease Financing Considerations. There are many variations on project lease structures. One example of a project lease financing structure requires that the site owner obtain temporary construction financing. On completion of construction, the lessor purchases the project for some agreed-on price, with the project revenues committed to repaying the lease payments. Generally, the lease terms will also contain provisions intended to protect the lessor's investment by providing, for example, for the establishment of special funds for capital repairs and replacements during the term of the lease. Any excess project benefits may be shared on some agreed-on formula with the site owner or taken in full by either party.

The merits of lease financing vary according to the relationships and responsibilities of the respective parties. In general, advantages include

- Transfer of tax benefits from lessee to lessor (capital leases),
- Nonrecourse debt for lessor (leveraged leases),
- Off-balance sheet accounting (operating leases),
- Lower interest (versus conventional bank financing),
- Flexible structure for joint ventures,
- Fixed-fate payment stream, and
- One-hundred percent financing available.

Disadvantages include

- Loss of residual (operating leases, when asset has substantial useful life on expiration of lease term),
- Cost higher (when tax benefits could have been used by lessee), and
- Fixed senior obligation against project.

PRIVATIZATION. Privatization entails development, ownership, and operation and maintenance of the electric-generation facility by an independent developer, and sale of output from the facility to the WWTP. For example, the WWTP may be unwilling to undertake development of an electric-generation plant because of high risk, capital costs, and/or other concerns. An outside developer may be obtained to evaluate and construct a suitable generation facility that meets the WWTPs power requirements, is compatible with WWTP operations, and produces sufficient economic returns to the developer.

JOINT OWNERSHIP AND/OR DEVELOPMENT. Between sole development and ownership by the WWTP owner and privatization, there are a wide variety of project structures that can be employed, depending on the extent of financial risk and rewards assumed by the respective project participants. Key negotiating points include

- Economic-value-attributed digester gas provided by the WWTP to the project as fuel;
- Economic-value-attributed use of the project site (that is, rental or lease charge); and
- Ownership of project equipment.

Negotiable contract points to consider include

- Commitment to specific levels of service, for example,
  - Thermal quantity and quality,
  - Electric quantity and quality,
  - Timing of delivery of service, and
  - Hookup provisions.
- Method of compensation to project participants, for example,
  - Independent developer assumes development risk, operating risk, and cost of capital and charges WWTP for power taken on some discounted basis.
  - Wastewater treatment plant owner and independent developer share development costs, risks, and benefits on some basis (generally proportional to equity contribution).

SHARED SAVINGS. One common mechanism offered by energy management companies, but also available through some cogeneration equipment manufacturers, is referred to as *shared savings.* Under the shared savings approach, the energy management company finances and implements energy conservation measures at no risk to the energy user. The energy management company takes its compensation in the form of shared savings; that is, the cost of implementing these energy conservation measures plus some component for return on investment is funded by the reduced cost of utility power and fuel purchases.

Many variations on this structure may be available:

- The energy management company might be paid a fixed annual management fee.
- The energy management company's share in energy cost savings may be subject to some minimum or maximum, either in magnitude or in number of years.
- The term for sharing savings may be subject to a fixed term, with or without prospect for renewal.

For a power-generation project, a large equipment manufacturer may offer to supply either the cogeneration unit or the entire generation plant on a turnkey basis with no up-front cost to the plant owner. Compensation to the manufacturer would then be structured on the basis of a stream of payments funded by the expected savings in purchased power costs.

ENERGY REQUIREMENTS PROFILE

Before embarking on a project, the WWTP owner should have an understanding of the WWTP's overall energy requirements, condition and type of equipment (classified by process function and power requirements), and fuel resource options.

Appendix Equations for Converting from English Units to Metric Units

atm × 101.3 = kPa

Btu × 1.055 = kJ

Btu/cu ft × 37.26 = kJ/m^3

Btu/gal × 278.7 = kJ/m^3

Btu/hr × 0.293 1 = W

Btu/lb × 2.326 = kJ/kg

cfm × (4.719×10^{-4}) = m^3/s

cfm/ft × 1.549 = L/m·s

cfs × (2.832×10^{-2}) = m^3/s

cu ft × (2.832×10^{-2}) = m^3

cu ft/lb × (6.243×10^{-2}) = m^3/kg

(°F – 32)0.555 6 = °C

gal × (3.785×10^{-3}) = m^3

gal × 3.785 = L

gph × (1.051×10^{-6}) = m^3/s

gpm × (6.308×10^{-5}) = m^3/s

in. × 25.40 = mm

in. × (2.540×10^{-2}) = m

ft × 0.304 8 = m

ft-lb × 1.356 = N·m

ft-lb/min × (2.259×10^{-2}) = W

ft-lb/sec × 1.355 = W

ft/sec × 0.304 8 = m/s

gal × (3.785×10^{-3}) = m^3

gpm × (6.308×10^{-5}) = m^3/s

gpm × 5.451 = m^3/d

hp × 745.7 = W

hp-hr × 2.685 = MJ

hp/mil. gal × 0.197 0 = W/m^3

in. × (2.540×10^{-2}) = m

in. Hg × 3.377 = kPa

kWh × 3.600 = MJ

kWh/d × 41.67 = W

kWh/lb × (7.936×10^{-3}) = MJ/kg

kWh/mil. gal × 951.1 = J/m^3

lb × 0.453 6 = kg

lb/cu ft × 16.02 = kg/m^3

lb/d × 5.250 = mg/s

lb/lb × 1 000 = g/kg

mgd × (4.383×10^{-2}) = m^3/s

mgd × (3.785×10^{3}) = m^3/d

mile × 1.609 = km

psi × 6 895 = Pa

scfm × (4.719×10^{-4}) = m^3/s

sq ft × (9.290×10^{-2}) = m^2

Index

A

B

C

D

F

G

H

I

L

M

N

O

P

R

S

T